Distributed Market-Grid Coupling
Using Model Predictive Control

In Fulfillment of the Requirements for the Degree of

Doctor of Engineering

Faculty of Informatics
Karlsruhe Institute of Technology (KIT)

Approved

Dissertation

by

Yong Ding

Born in Hangzhou, China

Examination Date: June 3rd, 2016
Primary Examiner: Prof. Dr.-Ing. Michael Beigl
Secondary Examiner: Prof. Dr.-Ing. Veit Hagenmeyer

Acknowledgements

First and foremost, I would like to express my gratitude to my primary advisor, Professor Dr. Michael Beigl, who has mentored me throughout my doctoral studies. His advice on both my research and career path have been tremendous. I would also like to thank Professor Dr. Veit Hagenmeyer for serving as my secondary advisor during my doctoral studies. It was an honor and a great pleasure to work with him and to be co-advised by him. His constructive comments and suggestions provided me with a lot of help for revising my dissertation. The research group TECO has been a source of friendships as well as good advice and collaboration. I am especially grateful for a great number of insightful and inspirational discussions with my colleagues and friends at TECO. A special thank you goes to our Director of Research Dr. Till Riedel and my officemate Mr. Martin Alexander Neumann, who were always available and helpful for help and discussion. I very much appreciated their enthusiasm and willingness to engage in frequent research discussions with me. All past and present Bachelor, Master and Diploma students that I have had the pleasure of working with, have helped me along the way with experiments and implementations. I would especially like to mention Ömer Kehri, Axel Morawietz and Erwin Stamm. I further would like to thank my friend Dr. Hossein Miri, who provided me with a great help to proof-read my dissertation.

Writing a dissertation is hardly possible without the family support. Words cannot express how grateful I am to my parents for all of the sacrifices that they have made for me. I thank them for their endless support and understanding throughout all the years, even if it meant not seeing me for more than 10 years. Last but not least, I would like to express my heartfelt appreciation and deepest gratitude to my beloved wife, Hua, whose love and encouragement have been the source of my power and inspiration, throughout our entire time together. She has spent sleepless nights with me discussing my work and providing fresh insights, as well as lent me her unconditional support in the moments where nobody else was there. Special thanks to her for putting up with my weekends in the office, for listening to my complaints, and for giving me the motivation to finish this dissertation. On countless occasions, I learned that "if at first you don't succeed, go back and do what your wife told you to do in the first place." And of course, to our two lovely daughters, Mia and Olivia, who brought me infinite joy and happiness in the stressful years of my Ph.D. pursuit.

Abstract

Real-time monitoring of electricity grids' power flow, which reflects the physical reality of the power system, plays a crucial role in the power market, since the real-time market behavior often deviates from long-term market forecasts, due to unexpected supply-demand imbalances and the resulting price volatility. The real-time market results, in turn, have a major influence on the optimal dynamic economic dispatch of the power generation for stabilizing the power load in the grid. However, an appropriate market-grid coupling, in terms of a real-time interaction between the market and the grid, has not been designed to be available either from the grid network side or from the market structure side. In particular, in the context of Demand Response (DR), an incentive-driven load shedding or shifting for grid relief cannot be realized without an appropriate market-grid coupling. In this dissertation, a feedback control concept is proposed, designed and evaluated for modeling a market-grid coupling. The dissertation, also, addresses the research question of whether the market price as a feedback signal can effectively control the power dispatch in the grid, and vice versa.

Previous research works on interactions between market dynamics and grid dynamics have mainly addressed market power analysis, oligopolistic competition modeling and stability analysis. Recently, researchers have focused primarily on investigations of a complex interaction between the market and the grid, in terms of interoperability or controllability. This dissertation addresses rather a combination of both interoperability and controllability; namely an interoperable control between the market and the grid, by means of a closed-loop feedback control system. In order to demonstrate that a MPC-based (Model Predictive Control) closed-loop feedback control system can be developed for this purpose, in terms of an interoperable control of market prices and power dispatch, we define three theses that are validated in this dissertation:

1. It is possible to model a system, in which both market and grid dynamics are mutually influenced by each other.

2. It is possible to develop a dynamic equilibrium model of distributed power grids with local power markets through a distributed MPC strategy.

3. Power load represents the link between the market and the grid, therefore an accurate load forecasting is beneficial for the MPC task.

Essentially, this dissertation presents a novel approach with a closed-loop feedback control concept for the distributed market-grid coupling. One important part of the main contribution of this dissertation is the formal definition of the market-grid coupling. As the first requirement for the market-grid coupling, a real-time market model is designed and formulated as a power balancing option; Subsequently, a two-layer grid model is presented for an optimal dynamic dispatch (ODD) study. Based on both models, a definition of the market-grid coupling is formalized with a feedback control loop. Then, a further investigation and analysis of this formalized market-grid coupling is conducted in two different directions. A co-simulation framework that realizes a market-grid coupling is developed for studying the grid load's influence on the market price. In order to extend this unidirectional control towards an interoperable control between the market and the grid within the market-grid coupling framework, the system modeling of a MPC-based closed-loop feedback control system is presented, in which a market price optimization and a power dispatch optimization are performed concurrently. The problem formulation of the control system firstly focuses on a coupling model with a single grid unit and its correspondent local market. Subsequently, a distributed control architecture by means of a hierarchical MAS (Multi-Agent System) is presented for extending the centralized MPC problem of a local market-grid coupling to a distributed MPC problem of a distributed market-grid coupling. A distributed MPC strategy is adopted to decompose the overall grid into interconnected grid units, so that individual grid units achieve control objectives collaboratively. Different valuation use cases with IEEE bus system test cases are introduced. Simulation-based numerical results show that not only the centralized MPC formulation, but also the distributed MPC formulation, provide a clear stability of both the market price and the power load dispatch. Finally, an adaptive load forecasting framework is proposed to improve the STLF (Short-Term Load Forecasting) performance. The obtained accurate load forecasting result shows its benefit for solving the above MPC problems.

The contributions of this dissertation include the following:

1. Identification and characterization of an interoperable control between the power market and the power grid;

2. Design of a closed-loop MPC for the market-grid coupling;

3. Extension of the single control loop with a collaborative distributed MPC strategy for coupling distributed markets and grids;

4. Development of an adaptive load forecasting framework.

Deutsche Zusammenfassung

Die Echtzeit-Überwachung des Energieflusses spielt eine entscheidende Rolle im Strommarkt. Denn aufgrund unerwarteter Angebot-Nachfrage-Ungleichgewichte und der daraus resultierenden Preisvolalität weicht das Marktverhalten oft von langfristigen Marktprognosen ab. Die Echtzeit-Marktergebnisse haben wiederum einen großen Einfluss auf die Stabilisierung des Energiesystems im Sinne von einer wirtschaftlichen optimalen Stromübertragung und -verteilung im physikalischen System. Eine geeignete Markt-Netz-Kopplung in Form einer Echtzeit-Interaktion zwischen dem Markt und dem Netz, steht dafür jedoch noch nicht zur Verfügung. Vor allem im Rahmen des sogenannten Demand Response (DR), kann ein anreizgetriebener Lastabwurf oder eine anreizgetriebene Lastverlagerung zur Netzentlastung nicht ohne eine entsprechende Markt-Netz-Kopplung realisiert bzw. optimiert werden. In dieser Dissertation wird daher ein Regelungskonzept vorgeschlagen und entwickelt, um die Modellierung einer Markt-Netz-Kopplung zu ermöglichen und zu evaluieren. Die Dissertation befasst sich auch mit der Fragestellung, ob der Marktpreis als eine Rückkopplungsvariable zur Regelung der Stromübertragung und -verteilung (eng. *power dispatch*) im Netz eingesetzt werden kann, und umgekehrt.

Die bisherigen Forschungsarbeiten über Interaktionen zwischen der Markt- und Netzdynamik beziehen sich hauptsächlich auf Analyse der Marktmacht, Modellierung des oligopolistischen Wettbewerbs und Stabilitätsanalyse. In letzter Zeit haben sich Forscher verstärkt mit Untersuchungen einer komplexen Interaktion zwischen dem Markt und dem Netz in Bezug auf die Interoperabilität bzw. die Steuerbarkeit/Regelbarkeit beschäftigt. Diese Dissertation befasst sich mit einer Kombination von der Interoperabilität und der Regelbarkeit, nämlich einer interoperable Regelung zwischen dem Markt und dem Netz mittels des Konzeptes eines geschlossenen Regelkreises. Um zu zeigen, dass sich im Hinblick auf eine interoperable Regelung der Marktpreise sowie Stromübertragung und -verteilung ein MPC-basierter (Model Predictive Control) Regelkreis für diesen Zweck eignet werden drei Thesen definiert, die in dieser Dissertation validiert werden:

1. Es ist möglich, ein System zu modellieren, in dem der Markt und das Netz hinsichtlich ihrer Dynamik inhärent wechselseitig voneinander beeinflusst werden.

2. Es ist möglich, ein dynamisches Gleichgewichtsmodell zwischen verteilten Stromnetzen und verteilten Strommärkten mittels einer verteilten MPC-Strategie zu entwickeln.

3. Der Stromfluss/Lastfluss stellt das Bindeglied zwischen dem Markt und dem Netz dar. Eine genaue Lastprognose ist daher für die MPC-basierte Lastoptimierung von Vorteil.

Im Wesentlichen präsentiert diese Dissertation einen neuartigen Ansatz mit einem Regelungskonzept eines geschlossenen Regelkreises für eine verteilte Markt-Netz-Kopplung. Ein wichtiger Beitrag dieser Dissertation ist die formale Definition der Markt-Netz-Kopplung. Als die Grundvoraussetzung für die Markt-Netz-Kopplung wird ein Echtzeit-Marktmodell als eine Option der Regelenergie konzipiert und formuliert, und gleichzeitig wird ein 2-Schichten-Netzmodell für eine optimale dynamische Stromübertragung und -verteilung vorgestellt. Darauf basierend wird eine formale Definition der Markt-Netz-Kopplung mittels eines Regelkreises eingeführt. Die Analyse dieser Markt-Netz-Kopplung wird in zwei Richtungen betrachtet. Zum einen wird ein Co-Simulations-Framework für die Untersuchung einer unidirektionalen Regelung hinsichtlich des Netzlast-Einflusses auf den Marktpreis entwickelt. Zum anderen wird die Systemmodellierung eines MPC-basierten Regelkreises vorgestellt, um eine interoperable Regelung zwischen dem Markt und dem Netz im Rahmen der Markt-Netz-Kopplung zu ermöglichen. Somit wird eine Optimierung des Marktpreises sowie Stromübertragung und -verteilung gleichzeitig durchgeführt. Die Problemformulierung dieses Regelungssystems konzentriert sich zunächst auf ein Kopplungsmodell zwischen einem einzelnen Ortsnetz und seinem entsprechenden Lokalmarkt. Anschließend wird eine verteilte Regelungsarchitektur mittels eines hierarchischen MAS (Multiagentensystem) vorgestellt, um das zentrale MPC-Problem einer lokalen Markt-Netz-Kopplung zu einem dezentralen MPC-Problem einer verteilten Markt-Netz-Kopplung auszubauen. Eine verteilte MPC-Strategie wird zur Spaltung des Gesamtnetzes in miteinander verbundene Ortsnetze verwendet, sodass die einzelnen Ortsnetze die Regelungsziele auf kollaborative Weise erreichen können. Verschiedene Evaluierungsszenarien mit Testfällen des IEEE-Bus-Systems werden eingeführt. Daraus ergebende numerische Ergebnisse zeigen, dass nicht nur der zentrale MPC-Ansatz, sondern auch der verteilte MPC-Ansatz, eine gute Stabilität sowohl des Marktpreises als auch der Stromverteilung bietet. Schließlich wird ein adaptives Lastprognose-Framework vorgeschlagen, um zu zeigen, dass eine genaue Lastprognose für die oben genannten MPC Probleme von Vorteil ist.

Contents

Contents

LIST OF TABLES

List of Figures

List of Figures

INTRODUCTION

We are facing a restructuring of the power industry towards a smart grid, which will be fully implemented in the Pan-European Grid Network with optimization foci on both physical grid operations and market designs [12]. In the course of smart grid development, one of its main features is the expansion of renewable power generation. Not only at the European level, one of the 2020's objectives has been defined towards at least 20% increase of renewable energy share, but also in Germany, through the EEG (Renewable Energy Law) policy for the further development of renewable energy, renewable power generation gains its dispatch priority with fixed feed-in tariffs regulated by law [6, 17]. The outcome of this development is an increasing share of Distributed Generation (DG) with highly intermittent energy resources of e.g. wind turbines and solar panels, which poses a challenge in terms of limited predictability and controllability [20]. Due to the merit order effect [32], even though intermittent renewable power generation could decrease the market price [5, 17], the increased volatility of renewables without sufficient storage capacity will lead to critical network loads of the power grid [34, 7]. In order to develop a smart-grid-ready power market with more and more penetration of intermittent renewables, we need to analyze the current market model and decide what kind of pricing mechanism to put in place. We, also, need to determine how to expand grid is the right choice for the smart grid.

In spite of various concepts such as a virtual power plant or a demand response program to cope with the volatility of renewables, it is reasonable to feed the renewable power generation into the grid for a long-term application, only if a market coupling to the grid exists [25]. Neuhoff showed in an executive summary of the smart power market project [25] that the power network load can be controlled by different pricing mechanisms. However, as Palensky and Dietrich [27] stated that the currently-implemented monetary incentives like RTP (Real-Time Pricing) does not reflect real-time physical situations of the power grid, such as generator ramp rates, transmission congestion, etc., which leads to the fact that load shedding or shifting for grid relief can not be realized via market-based real-time pricing alone. Therefore, the research question here arises whether more information of the physical power system mapped onto prices, meaning a grid-driven real-time pricing mechanism, can bring DSO (Distribution System Operator) and TSO (Transmission System Operator) the advantage that the existing grid infrastructure is then exploited much better for a dynamic economic dispatch [38], thereby controlling consumer behavior specifically towards stability with appropriate quality criterion [21].

In order to address the above question, this dissertation concerns itself with a closed-loop feedback system that models a market-grid coupling in terms of the real-time interaction between the market and the grid. Real-time monitoring of the grid's power flow, which reflects the physical reality of the power system, plays a crucial role in the power market, since the real-time market behavior often deviates from long-term market forecasts due to more and more unexpected supply-demand imbalances [10] and the resulting price volatility [33, 36]. The real-time market results have in turn a major influence on the optimal dynamic economic dispatch of the power generation for stabilizing the power load in the grid [29]. Therefore, in addition to modeling the market-grid coupling, this dissertation focuses further on the investigation of market price as a feedback signal to control the dispatch schedule of generation units and vice versa, thereby demonstrating the capability of the proposed feedback control system for achieving a stable optimal dispatch [11].

1.1 MOTIVATION

The vision of the smart grid represents not only the creation of intelligent power supply networks to allow efficient and reliable use of energy resources, but also the redesign of the market structure coupled with it. In order to develop a smart-grid-ready power market, on the one hand the integration of the physical reality of the power grid — in particular the grid operational integration of renewables — into the economic market model has been set as the first requirement [19]; on the other hand progressively market-integrated solutions have been examined with their efficiency to the grid load balancing task [23]. Although the market-grid coupling is not a well established field, interactions between market dynamics and grid dynamics have already been taken into consideration in the following power system research subjects:

Market power analysis: In the early stage of the power market deregulation, researchers noticed that physical transmission constraints could have a significant impact on market power mitigation or regulation [31]. Therefore, Hogan [15] suggested a MPEC (mathematical program with equilibrium constraints) approach for the market power analysis subject to transmission constraints towards an "economic dispatch". By evaluating market power in congested power networks, Oren [26] demonstrated the importance of including transmission rights and conditions within competitive power market to prevent the price distortion.

Oligopolistic competition modeling: Transmission constraints not only have a significant impact on the market power regulation, but also play an important role for an oligopolistic market equilibrium. In the MPEC approach of Hobbs et al. [14] and the CSF (conjectured supply function) approach of Day et al. [8], an ISO (independent system operator) model was proposed to describe the impact of power flow constraints on market equilibrium prices, in which an efficient rationing of transmission capacity subject to power flow constraints of the grid network was determined. Considering physical constraints of the grid, an oligopolistic competition based market model for both power and spinning reserve (SR) could indicate an increasing influence on the power market efficiency with respect to the opportunity cost derived from power shifts between producing and spinning [4].

Stability analysis: The power system is a physical basis for the power market and meanwhile uses the power market as an operation platform for balancing. To study the stability of both components as well as the controllability between them, Xue [39] proposed an integrated physic-economic power system, which takes interactions between power market dynamics and power system stability on different time scales into account. Compared to the market stability without integration of power system dynamics, Alvarado et al. [2] examined the stability of an interconnected power market model with physical dynamics of a power system. The more accurate the coupled model is, the higher sensitivity of market price changes has been shown for maintaining the system stability with respect to power imbalance. Moreover, Liu and Wu [22] conducted a stability analysis on the power market equilibrium subject to transmission constraints, which showed that no Nash equilibrium could be maintained and the market equilibrium could exist iff on the constraint boundary with no nodal price difference or no congestion charge.

Interoperability or controllability: Demand-side management programs require pricing mechanisms towards dynamic or real-time pricing subject to the grid stability, for which researchers have begun to investigate the complex interactions between the market and the grid in terms of interoperability or controllability. To the best of our knowledge, Alvarado et al. [1, 2] and Widergren et al. [37] were the first authors who proposed a coupling concept for studying dynamic control interactions between the power market and the power system. The work of Alvarado et al. focused on the mathematical formulation and the stability analysis of the coupling concept, while the work of Widergren et al. focused on the requirements of the coupling concept towards a co-simulation. To study the balancing capability of supply and demand in such a coupled system, Roozbehani et al. [29] introduced an additional interaction between the wholesale and retail market. The proposed real-time retail pricing mechanism based on the wholesale locational marginal prices provided a control concept for stabilizing the coupled power system with demand response. Furthermore, Wang et al. [35] proposed a redesign of the market structure with a grid model extension by an accurate representation of the underlying technical characteristics and limitations of the power production and transmission facilities, in order to cope with the problem of high volatility of the market clearing price. Based on the coupled model, authors opened up a discussion about control capabilities of the market-driven grid model and vice versa.

This dissertation is inspired by the "coupling" concept proposed by Alvarado et al. and Wang et al. for an interoperable control between the power market and the power grid, in order to analyze and understand the capabilities and limitations of the employed feedback control concept for a market-grid coupling. The motivation and impetus is to further develop a feedback control system, in which individual market and grid units work collaboratively for maintaining the stability of a distributed market-grid coupling system.

1.2 GOALS AND SCOPE

An appropriate coupling model between the market and the grid is not only of particular importance for the above four research subjects, but also fundamental for Demand Response (DR) concerning challenges, such as 1) demand-side uncertainties in e.g. consumer response behavior and 2) stability and volatility of both price and demand [30]. Recently, researchers have introduced **Model Predictive Control** (MPC) as an efficient approach for this coupling, which is formulated mainly as an **Optimal Control Dynamic Dispatch** (OCDD) problem [38]. Examples include a game-theoretical MPC framework for the market price control using generators' ramp rates as a control variable [16] and a distributed MPC approach to deal with a combined environmental and economic dispatch problem [9]. From a control theoretical point of view, both approaches provide only an open-loop optimal solution to the market price control. The main issues of these open-loop OCDD problems are inaccuracies originating from modeling uncertainties and no guarantee of the dispatch system stability. In contrast, we aim at not only an optimal market price control but also a stable optimal dispatch. One promising approach is the closed-loop control mechanism by the MPC method, so that advantages of both long-term planning (feedforward price control based on grid performance predictions over a predefined time horizon) and reactive control (feedback dispatch control using price measurements as input) can be combined [28].

Therefore, **the main goal of this dissertation is twofold: 1) determination and formulation of the system models and the correspondent parameters for the market-grid coupling; 2) development of a MPC-based closed-loop feedback control system for the formalized market-grid coupling, in order to achieve optimal power dispatch control via market prices, and concurrently stabilize market prices via power re-dispatch.** Due to future large-scale power systems

5

with a high penetration of distributed generation units and complex constraints, **an ancillary goal of this dissertation, which provides almost equal scientific importance, is to adopt a distributed MPC strategy to decompose the overall grid into interconnected grid units, and subsequently achieve the above control objectives of individual grid units collaboratively.**

The objective and the main aspects of the proposed coupling system are demonstrated schematically in Figure 1. Following the local power market idea of Hatziargyriou et al. [13] and Menniti et al. [24] to optimize the operating cost within a microgrid, we propose firstly to disaggregate a complex grid network into several interconnected grid units (such as microgrids at different scales). Then, our proposed decoupled control loop will be applied in each grid unit to integrate the physical reality of the grid unit into the corresponding local market (and vice versa) for a dynamic equilibrium between the grid unit and the local market (see Figure 1a). At a high level of abstraction, our visionary proposal can be represented by an encapsulated market service, which is not only applicable to specific regions, but also by means of cascading the service platform applicable to the whole of the European market. For example, in a country like Germany, this service platform can be applied in a top-down approach to various states and regions, and equally to cities, offices and homes (see Figure 1b). **Two important limitations to the above defined goal are that we will in this work only consider 1) the same pricing mechanism for each local market and 2) one disaggregation layer for all grid units.** Different market pricing mechanisms at the same time or a cascade control system for the grid architecture with more than one disaggregation layer will increase the complexity of the coupling analysis, which is out of the scope of this dissertation and will be addressed as future work.

1.3 THESES OF THE DISSERTATION

This dissertation presents a novel approach with a closed-loop control concept for the distributed market-grid coupling, in order to demonstrate the interoperable controllability between distributed market and grid units for stabilizing power dispatch and market prices concurrently. It, also, puts forward the following conjectures and theses that are verified through a succession of numerical analysis and simulations.

<u>Thesis 1</u>: **It is possible to model a system, in which both market and grid dynamics are mutually influenced by each other.**

This thesis focuses on the general feasibility of a market-grid coupling through a closed-loop control concept. After the definition of our proposed coupling and the corresponding requirements analysis for an interoperable control in

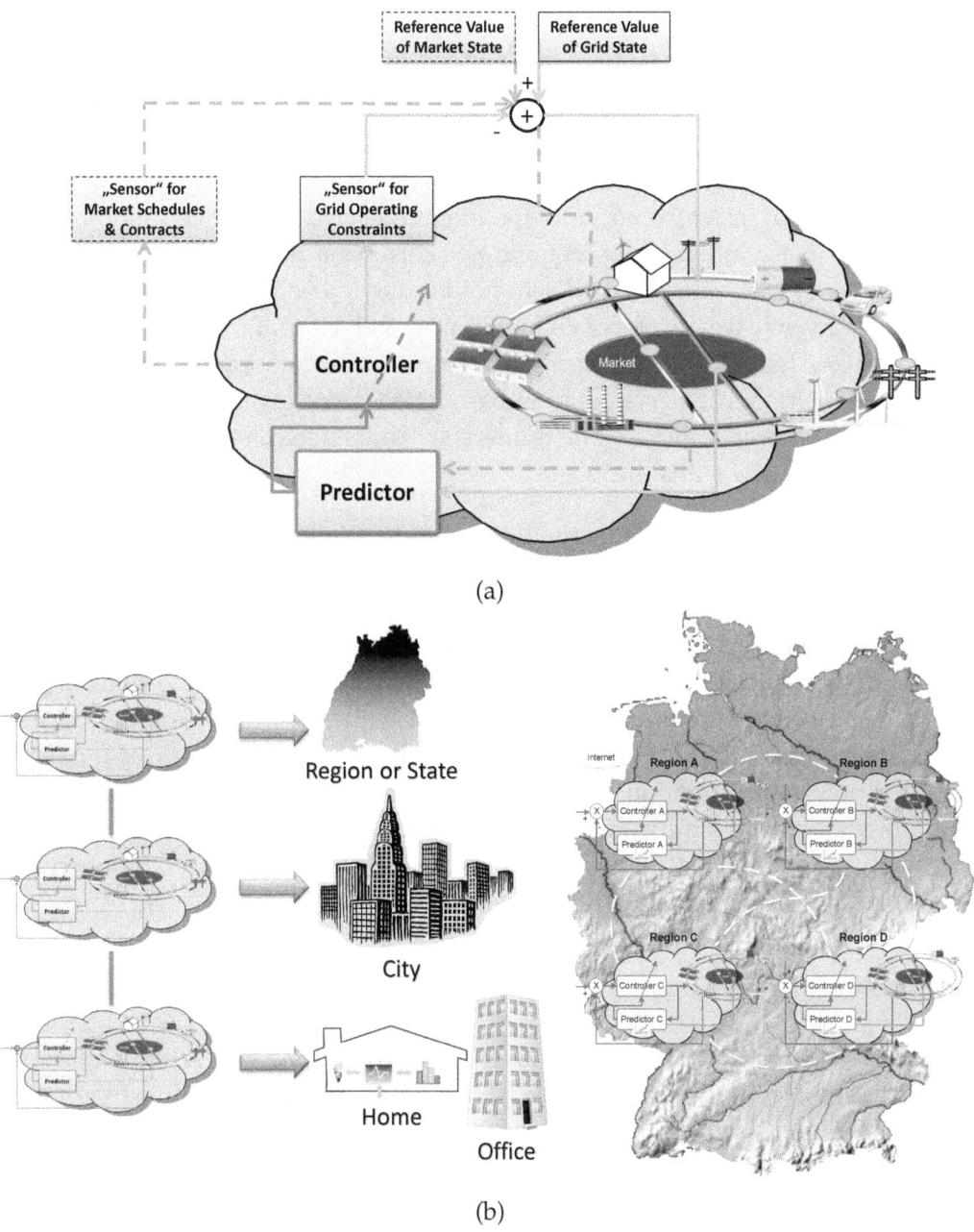

(a)

(b)

Figure 1.: The vision and concept of the distributed Market-Grid Coupling - (a) The vision for a local grid unit; (b) The vision for global distributed grid units

Chapter 3, it is demonstrated in Chapter 4 how market and grid dynamics can be modeled as the system plant for a control loop.

Thesis 2: **It is possible to develop a dynamic equilibrium model of distributed power grids with local power markets through a distributed MPC strategy.**
This thesis assumes that a distributed market-grid coupling system can be constructed, in which not only the interconnected grid units share information, but also individual grid units and corresponding local market units exchange output values, in order to hold a dynamic equilibrium among them. In Chapter 5, we develop a cooperation-based MPC algorithm with a multi-agent system (MAS) architecture to provide a collaborative capability. We implement then this cooperation-based MPC on a given grid network with one disaggregation layer. The numerical results in terms of a control stability analysis validate this thesis.

Thesis 3: **Power load represents the link between the market and the grid, therefore an accurate load forecasting is beneficial for the MPC task.**
The predicted load is the starting point for forecasting the load distribution in the power grid, which helps to calculate the predicted generator capacity and subsequently determine the power flow prediction in the transmission line based on the net power injection. In Chapter 6, we develop an adaptive load forecasting framework to analyze a bottom-up approach of modeling the power consumption at different granularity levels. Considering the load forecasting as an integrated process of the MPC task, a numerical study on power dispatch control with load forecasting shows the correlation between load forecasting accuracy and dispatch costs' decrease.

1.4 CONTRIBUTIONS

This work addresses the above-mentioned three theses and achieves the objective of a distributed market-grid coupling. The main contributions of this dissertation refer to 1) the identification and characterization of an interoperable control between the power market and the power grid; 2) a closed-loop MPC design for the market-grid coupling; 3) an extension of the single control loop with a collaborative distributed MPC strategy for coupling distributed markets and grids; 4) the development of an adaptive load forecasting framework, which are briefly described as follows:

Contribution 1: **Identification and characterization of an interoperable control between the power market and the power grid**
This contribution identifies the system variables of both market and grid,

which are used to model the system plant of a market-grid coupling. The proposed market-grid coupling focuses on an interoperable control aspect between the power load and the power price. We, first, model the market system and the grid system in a way that the output variable of either system can be taken into account as a control variable for the other. Based on the identified market and grid model, we, then, specify a feedback control loop for coupling both systems.

Contribution 2: Design of a closed-loop MPC for the market-grid coupling

This contribution demonstrates an extension of a MPC-based dispatch control towards a closed-loop feedback control system for both market price control and power dispatch regulation. Within each defined grid unit, we employ a stabilizing pricing mechanism based on the Locational Marginal Prices (LMPs) to determine stable local market prices. The LMPs are calculated based on the Optimal Power Flow (OPF) problem of the individual grid units. We assume that the local market prices directly influence consumers' power consumption behavior. Given value or utility functions of the consumers, we are able to use the local market prices as a feedback signal to control the dispatch schedule of generation units that hold the supply-demand balance and preserve the OPF constraints.

Contribution 3: Extension of the single control loop with a collaborative distributed MPC strategy for coupling distributed markets and grids

This contribution extends the single closed-loop MPC with a distributed MPC algorithm for coupling distributed power markets and power grids in real time. For market price control and power dispatch regulation on distributed grid units, we develop a distributed and collaborative control architecture by means of a MAS and focus on cooperation aspects among the distributed grid units in terms of information exchange, in order to provide a collaborative capability for suppressing constraints violation. In order to guarantee the closed-loop stability, the MPC agent of each grid unit exchanges with its neighboring grid units not only the information of state and input variables, but also the controller objectives. Through simulation-based numerical results, we show the capability of the proposed collaborative distributed MPC strategy for a stable and robust coupling.

Contribution 4: Development of an adaptive load forecasting framework

This contribution concerns the development of an adaptive load forecasting framework, which can be incorporated in the MPC-based feedback control loop as an integrated process for a more cost-efficient load distribution. We focus on a framework for implementing Short-Term Load Forecasting (STLF), in which statistical time series prediction methods and machine learning-based regression methods can be configured to benchmark their performance against each other. In addition, our framework incorporates two features regarding pre-processing and prediction modeling: 1) it wavelet transforms the training data during the pre-processing phase to better extract redundant information from power load data; 2) it integrates activity information as an additional load influencing factor which reflects activity-driven load patterns. Studies on real-world data at different granularity levels show that activity information as an influencing factor could improve the STLF accuracy to varying degrees and the aggregation level of power load data and activity data matters. By incorporating the load forecasting results into the MPC problem formulation, a further cost reduction and stabilization for the power re-dispatch can be demonstrated based on a numerical study.

1.5 DISSERTATION STRUCTURE

The structure of this dissertation, seen in Figure 2 schematically, is divided as follows: Chapter 1 introduces the motivation and the goal of this work, and gives an overview of the core theses as well as the main contributions. Chapter 2 describes the relevant related work to this dissertation. The remainder of this dissertation, which presents the details of the main contributions, is divided as follows: Chapter 3 presents a power market modeling and a power grid modeling that are employed in this dissertation, and based on that provides a formal definition of the proposed market-grid coupling. Technical requirements for implementing the defined market-grid coupling concept in terms of an interoperable control is followed as well in this chapter. Chapter 4 illustrates the system modeling of a MPC-based closed-loop feedback control system for an interoperable control between the market and the grid, in which a market price optimization and a power dispatch optimization are performed concurrently. The problem formulation of the control system focuses on a coupling model with a local grid unit and its correspondent local market. Chapter 5 extends the in Chapter 4 proposed feedback control framework with a distributed MPC approach. By means of MAS, the centralized coupling problem for a local grid unit is developed towards a distributed coupling problem for distributed grid units. Chapter 6 introduces a STLF framework, in which statistical time series prediction methods and machine learning-based regression methods, can be configured to benchmark their performance against each other on given datasets of power consumption and other related exogenous variables. To investigate the capability of the proposed framework, this chapter presents 3 case studies with real-world datasets and an integration test in the MPC problem formulation of Chapter 5. Chapter 3 to Chapter 6 all analyze experimental results based on numerical simulations, respectively. Finally, Chapter 7 concludes the dissertation with a summary and some closing remarks as well as potential future research directions.

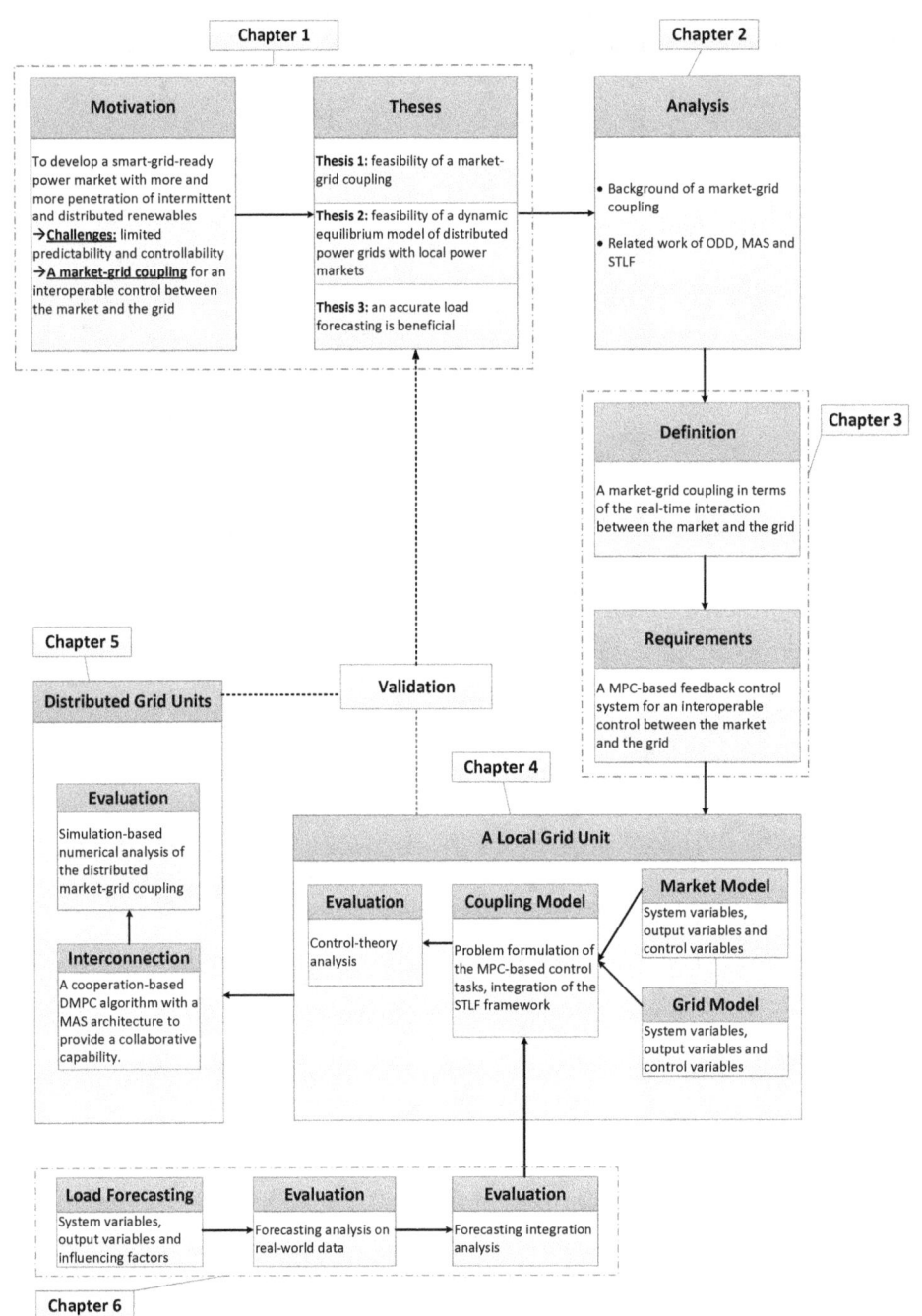

Figure 2.: Composition and structure of the dissertation

BIBLIOGRAPHY

[1] F.L. Alvarado, J. Meng, W.S. Mota, and C.L. DerMarco. Dynamic coupling between power markets and power systems. In *Power Engineering Society Summer Meeting, 2000. IEEE*, volume 4, pages 2201–2205 vol. 4, 2000. doi: 10.1109/PESS.2000.866987.

[2] F.L. Alvarado, J. Meng, C.L. DeMarco, and W.S. Mota. Stability analysis of interconnected power systems coupled with market dynamics. *Power Systems, IEEE Transactions on*, 16(4):695–701, Nov 2001. doi: 10.1109/59.962415.

[3] G. Andersson. Modelling and analysis of electric power systems. Technical report, ETH Zurich, 2008.

[4] G. Bautista, V.H. Quintana, and J.A. Aguado. An oligopolistic model of an integrated market for energy and spinning reserve. *Power Systems, IEEE Transactions on*, 21(1):132–142, Feb 2006. doi: 10.1109/TPWRS.2005.860909.

[5] Sven Bode and Helmuth-M. Groscurth. The impact of pv on the german power market. *Zeitschrift fuer Energiewirtschaft*, 35(2):105–115, 2011. ISSN 0343-5377. doi: 10.1007/s12398-010-0041-x.

[6] Bundesnetzagentur. Evaluierungsbericht zur ausgleichsmechanismusverordnung. Technical report, Federal Network Agency, Bonn, 2012.

[7] E.J. Coster, J.M.A. Myrzik, B. Kruimer, and W.L. Kling. Integration issues of distributed generation in distribution grids. *Proceedings of the IEEE*, 99(1): 28–39, Jan 2011.

[8] C.J. Day, B.F. Hobbs, and J.-S. Pang. Oligopolistic competition in power networks: a conjectured supply function approach. *Power Systems, IEEE Transactions on*, 17(3):597–607, Aug 2002. doi: 10.1109/TPWRS.2002.800900.

[9] Alejandro J. del Real, Alicia Arce, and Carlos Bordons. Combined environmental and economic dispatch of smart grids using distributed model predictive control. *International Journal of Electrical Power & Energy Systems*, 54(0):65 – 76, 2014. ISSN 0142-0615. doi: http://dx.doi.org/10.1016/j.ijepes.2013.06.035.

[10] Yong Ding, M.A. Neumann, M. Budde, M. Beigl, P.G. Da Silva, and Lin Zhang. A control loop approach for integrating the future decentralized power markets and grids. In *Smart Grid Communications (SmartGridComm),*

2013 IEEE International Conference on, pages 588–593, Oct 2013. doi: 10.1109/ SmartGridComm.2013.6688022.

[11] Yong Ding, Sozo Inoue, Martin A. Neumann, Erwin Stamm, Xincheng Pan, and Michael Beigl. A Personalized Load Forecasting Enhanced by Activity Information. In *1st IEEE International Smart Cities Conference (ISC2-2015)*, Guadalajara, 2015. IEEE.

[12] EEGI. The eegi research and innovation roadmap 2013-2022. Technical report, European Electricity Grid Initiative, 2013.

[13] ND Hatziargyriou, A Dimeas, AG Tsikalakis, JA Pecas Lopes, G Karniotakis, and J Oyarzabal. Management of microgrids in market environment. In *Future Power Systems, 2005 International Conference on*, pages 7–pp. IEEE, 2005.

[14] B.F. Hobbs, C.B. Metzler, and J.-S. Pang. Strategic gaming analysis for electric power systems: an mpec approach. *Power Systems, IEEE Transactions on*, 15(2): 638–645, May 2000. doi: 10.1109/59.867153.

[15] William W Hogan. A market power model with strategic interaction in electricity networks. *The Energy Journal*, pages 107–141, 1997.

[16] A. Kannan and V.M. Zavala. A game-theoretical model predictive control framework for electricity markets. In *Communication, Control, and Computing (Allerton), 2011 49th Annual Allerton Conference on*, pages 1280–1285, Sept 2011. doi: 10.1109/Allerton.2011.6120315.

[17] Janina C. Ketterer. The impact of wind power generation on the electricity price in germany. *Energy Economics*, 44(0):270 – 280, 2014. ISSN 0140-9883. doi: http://dx.doi.org/10.1016/j.eneco.2014.04.003.

[18] Hassan K Khalil and JW Grizzle. *Nonlinear systems*, volume 3. Prentice hall New Jersey, 1996.

[19] Corinna Klessmann, Christian Nabe, and Karsten Burges. Pros and cons of exposing renewables to electricity market risks - a comparison of the market integration approaches in germany, spain, and the {UK}. *Energy Policy*, 36(10):3646 – 3661, 2008. ISSN 0301-4215. doi: http://dx.doi.org/10.1016/ j.enpol.2008.06.022. URL http://www.sciencedirect.com/science/ article/pii/S0301421508002978.

[20] S. Knab, K. Strunz, and H. Lehmann. Smart grid: The central nervous system for power supply. Technical report, Universittsverlag der TU Berlin, 2010.

[21] Na Li, Lijun Chen, and S.H. Low. Optimal demand response based on utility maximization in power networks. In *Power and Energy Society General Meeting, 2011 IEEE*, pages 1–8, July 2011. doi: 10.1109/PES.2011.6039082.

[22] Youfei Liu and F.F. Wu. Impacts of network constraints on electricity market equilibrium. *Power Systems, IEEE Transactions on*, 22(1):126–135, Feb 2007. doi: 10.1109/TPWRS.2006.889083.

[23] Henrik Lund, Anders N. Andersen, Poul Alberg stergaard, Brian Vad Mathiesen, and David Connolly. From electricity smart grids to smart energy systems – a market operation based approach and understanding. *Energy*, 42(1):96 – 102, 2012. ISSN 0360-5442. doi: http://dx.doi.org/10.1016/j.energy.2012.04.003. URL `http://www.sciencedirect.com/science/article/pii/S0360544212002836`. 8th World Energy System Conference, {WESC} 2010.

[24] D. Menniti, Anna Pinnarelli, and N. Sorrentino. Operation of decentralized electricity markets in microgrids. In *Electricity Distribution - Part 1, 2009. CIRED 2009. 20th International Conference and Exhibition on*, pages 1–4, 2009.

[25] Karsten Neuhoff. European smart power market project - executive summary. Technical report, Climate Policy Initiative / DIW Berlin, 2011.

[26] Shmuel S Oren. Economic inefficiency of passive transmission rights in congested electricity systems with competitive generation. *The Energy Journal*, pages 63–83, 1997.

[27] P. Palensky and D. Dietrich. Demand side management: Demand response, intelligent energy systems, and smart loads. *Industrial Informatics, IEEE Transactions on*, 7(3):381–388, Aug 2011. doi: 10.1109/TII.2011.2158841.

[28] J.B. Rawlings. Tutorial overview of model predictive control. *Control Systems, IEEE*, 20(3):38–52, Jun 2000. ISSN 1066-033X. doi: 10.1109/37.845037.

[29] M. Roozbehani, Munther Dahleh, and S. Mitter. Dynamic pricing and stabilization of supply and demand in modern electric power grids. In *Smart Grid Communications (SmartGridComm), 2010 First IEEE International Conference on*, pages 543–548, Oct 2010. doi: 10.1109/SMARTGRID.2010.5621994.

[30] M. Roozbehani, M.A. Dahleh, and S.K. Mitter. Volatility of power grids under real-time pricing. *Power Systems, IEEE Transactions on*, 27(4):1926–1940, Nov 2012. doi: 10.1109/TPWRS.2012.2195037.

[31] R.E. Schuler. Analytic and experimentally-derived estimates of market power in deregulated electricity systems: Policy implications for the management and institutional evolution of the industry. In *Systems Sciences, 1999. HICSS-32. Proceedings of the 32nd Annual Hawaii International Conference on*, volume Track3, pages 10 pp.–, Jan 1999. doi: 10.1109/HICSS.1999.772840.

[32] Frank Sensfuss, Mario Ragwitz, and Massimo Genoese. The merit-order effect: A detailed analysis of the price effect of renewable electricity generation on spot market prices in germany. *Energy Policy*, 36(8):3086 – 3094, 2008. ISSN 0301-4215. doi: http://dx.doi.org/10.1016/j.enpol.2008.03.035.

[33] G. Verbic and C.A. Canizares. Probabilistic optimal power flow in electricity markets based on a two-point estimate method. *Power Systems, IEEE Transactions on*, 21(4):1883–1893, 2006.

[34] P.N. Vovos, A.E. Kiprakis, A.R. Wallace, and G.P. Harrison. Centralized and distributed voltage control: Impact on distributed generation penetration. *Power Systems, IEEE Transactions on*, 22(1):476–483, Feb 2007.

[35] G. Wang, A. Kowli, M. Negrete-Pincetic, E. Shafieepoorfard, and S. Meyn. A control theorist's perspective on dynamic competitive equilibria in electricity markets. In *Proc. 18th World Congress of the International Federation of Automatic Control (IFAC)*, Milano, Italy, 2011.

[36] G. Wang, M. Negrete-Pincetic, A. Kowli, E. Shafieepoorfard, S. Meyn, and U. Shanbhag. Dynamic competitive equilibria in electricity markets. In A. Chakrabortty and M. Illic, editors, *Control and Optimization Theory for Electric Smart Grids*. Springer, 2011.

[37] S.E. Widergren, J.M. Roop, R.T. Guttromson, and Z. Huang. Simulating the dynamic coupling of market and physical system operations. In *Power Engineering Society General Meeting, 2004. IEEE*, pages 748–753 Vol.1, June 2004. doi: 10.1109/PES.2004.1372914.

[38] X. Xia and A.M. Elaiw. Optimal dynamic economic dispatch of generation: A review. *Electric Power Systems Research*, 80(8):975 – 986, 2010. ISSN 0378-7796. doi: http://dx.doi.org/10.1016/j.epsr.2009.12.012.

[39] Yusheng Xue. Interactions between power market stability and power system stability [j]. *Automation of Electric Power Systems*, 21:7, 2002.

2 | RELATED WORK

This chapter provides a brief overview of the related work of this dissertation. Since the dissertation work is concerned with a distributed market-grid coupling system based on a feedback control concept, which refers to the development of individual system components in the control loop as well as a distributed control architecture, the related work focuses on the methods and technologies for enabling this development. Our work builds on previous works mainly in three different research fields: **Optimal Dynamic Dispatch (ODD), Multi-Agent Systems (MAS)** and **Short-Term Load Forecasting (STLF)**.

2.1 OPTIMAL DYNAMIC DISPATCH FOR ANALYZING A MARKET-GRID COUPLING

One of the key research questions that this dissertation addresses refers to optimal dynamic dispatch (ODD) problems in power systems, which were firstly introduced by Bechert and Kwatny [7] as an optimal control problem. In general, the ODD problems concerning the formulation can be classified into either optimization theory based DED (dynamic economic dispatch) or control theory based OCDD (optimal control dynamic dispatch) [98]. Both DED and OCDD have a similar objective towards determining the optimal operation schedule of the committed generation units, so as to balance the predicted time-varying load demand over a time horizon satisfying various system and operational constraints. From a control-theoretical point of view, Xia et al. [99, 98] stated that both OCDD and DED mainly addressed in fact open-loop dispatch problems.

In order to compensate inaccuracies originating from modeling uncertainties of ODD problems and thereby to maintain the stability of the dispatch system, several recent research work focused on closed-loop control mechanisms based on MPC, which is also one of the main foci of this dissertation. Xia et al. [99, 100] presented an extended DED problem that incorporated DED and OCDD taking ramp rate violations into account, and proposed MPC as the solution that took average ramp rates at each dispatch interval as a feedback signal to control future generation output schedules. Elaiw et al. applied the same MPC approach as in the work of Xia et al. [99, 100] for a profit-based objective function that took the trade off between generation and emission into account [26] and for an extended fuel cost minimization function that determined the optimal dispatch of combined heat and power co-generation [27]. Xie et al. [102] proposed a MPC-based dispatch algorithm to minimize both generation and environmental costs, which used the expected total load as well as the expected maximum output levels of intermittent resources as a feedback signal to control generation output levels. Another MPC-based research work by Biegel et al. [9] focused on consumers with storage capacity, and considered required power generation and consumers' storage rates as a feedback signal to control the load balance.

Not only a standard MPC-based closed-loop approach has shown its potential for solving the ODD problems, but also some hybrid feedback control mechanisms have proven their value in this context. Jokic et al. [48] formulated their OCDD problem as an optimal power flow problem with congestion constraints in transmission networks and presented a MPC-based hybrid control scheme for achieving economically optimal power balancing and congestion control. In the proposed hybrid feedback control loop, an explicit piecewise affine dynamic controller used nodal prices as a feedback signal to control the network frequency deviation and the power line overload, while a MPC was responsible for the limitation of the frequency deviation and the power line overload during the transient period. Liang et al. [55] used recurrent neural networks (RNNs) to design a dynamic OPF controller for tracking a power systems optimal operating point in terms of an OPF solution. This closed-loop control proposal employed the OPF-based generation adjustments as a feedback signal to simultaneously control the network frequency, the transmission line power flow and the generator terminal voltage. A scenario-based stochastic MPC algorithm is proposed by Patrinos et al. [74] for the market-driven optimal power dispatch problem using both market price and generation schedule as a feedback signal. Compared to the classical MPC-based ODD formulation, authors established a stochastic setting for the MPC problem due to stochastic processes of the intermittent power generation units, where a scenario tree generation algorithm is used for modeling the stochastic processes of the generation intermittency. Recently, some ODD research work [36, 19, 83] focused on a multi-level control architecture of the MPC-based approach for hybrid power generation and storage units. As one example of such work, Sachs et al. [83] proposed a two-layer MPC optimization approach to calculate the optimal power dispatch for a microgrid consisting of a PV, a battery and a diesel generator. In order to achieve the optimization objective of generator cost minimization, battery usage optimization as well as renewables usage maximization, authors presented a two-layer control architecture with respect to the robustness towards prediction errors. A main optimization layer is formulated as a MPC problem for the optimal control trajectory of both diesel power generation and battery SOC (state of charge), while a sub-optimization layer is formulated as a BVP (boundary value problem) for optimizing the diesel generator turn on/off time through the minimization of the SOC control error determined within the MPC layer.

Our work extends above proposed feedback control loops with a market module, in order to study the impact of optimal dynamic dispatch on the market price stability as well as the importance of dynamic pricing for achieving a steady-state stability of the grid. Recently, there have been several research work towards market-oriented power dispatch [104, 59, 50, 92, 40], which studied mainly the impact of intermittent energy resources, thereby focusing more on generation and storage dispatch, considering less power flow dispatch for a grid's steady-state stability. Furthermore, our proposed distributed control architecture, realized by means of a multi-agent system, distinguishes our work from the above-mentioned research works that mainly apply a centralized MPC architecture. Applications of distributed MPC techniques have already proven their effectiveness in solving large-scale dispatch problems for e.g. network overload control [42, 51], generation control [89, 63, 18] and load balance control [39, 18].

2.2 MULTI-AGENT SYSTEM AS A DISTRIBUTED CONTROL ARCHITECTURE

In recent years, multi-agent systems (MAS) [32] have been proposed as tools to implement various power system and market applications [60, 34], such as distributed energy control and management systems in (micro)grids [33, 81] and market behavior simulation as a complex adaptive system [17, 62]. In particular in the power dispatch field, MAS could also prove themselves for their applications as a distributed control approach. In the early stage, as stated by Yu and Liu [103] as well as Roche et al. [81], most MAS-based dispatch algorithms are market-based implementation of trading and clearing algorithms. They focused mainly on optimizing matches between load and generator bids for either cost minimization or profit maximization as the dispatch objective, which means the dispatch perspective in the sense of grid optimization was not considered. First in the work of Nagata et al. [65, 67], the authors proposed a MAS approach as a power re-dispatch formulation to solve the power system restoration. Afterwards, more and more contributions with MAS as a power dispatch algorithm have been made for a range of power system applications, including microgrid operation control [47, 57, 15, 16, 79], grid relief [24, 88, 80], energy resource scheduling [56, 58], generation/grid expansion [11, 35], voltage regulation [5, 15], fault diagnostics [61], etc.

The focus of the MAS in this dissertation is on realizing a distributed control architecture for a distributed market-grid coupling based on the power dispatch. Due to the distributed nature of distributed generators (DGs), there have been lots of MAS research contributions towards distributed architectures for the DGs' power dispatch control. Baran et al. [5] proposed a MAS-based dispatch strategy to optimize the voltage regulation on distribution feeders with DGs in a distributed manner. Authors formulated the voltage regulation with DGs as a reactive power dispatch problem, in which each DG agent is primarily mainly responsible for its own reactive power dispatch and voltage monitoring on that local DG node, and provides its dispatch capability as a bid. A predefined Moderator agent (one of the DG agents) decides based on the received bids incrementally, on which DG node and to what extend its reactive power dispatch should be performed, in order to achieve a necessary voltage support on the global feeder. Also in the context of microgrids, Bidram et al. [8] developed a MAS-based cooperative control scheme for the secondary voltage and frequency control. The same as in the work of Baran et al. [5], authors here proposed a distributed control architecture as well, in which each DG agent aims at minimizing deviations between the terminal voltage/frequency and the corresponding reference values on that DG node. Compared to the sequential cooperation scheme in the work of Baran et al. [5], Bidram et al. [8] presented rather a parallel cooperation scheme, since each DG agent only requires the information of its own and neighbors in the communication range for making a local control decision. Moreover, Pantoja and Quijano [73] employed a hierarchical MAS-based control scheme proposed by Dimeas and Hatziargyriou[20] to implement the replicator dynamics (RD) strategy for DG resources allocation in a power distribution system with microgrids. In spite of the distributed control strategy on each local DG node, a central or hierarchical coordination through the microgrid central controllers (MGCC) could not exclude any single point of failure and may have scalability problems.

For our market-grid coupling model, we take into account not only the generations units, but also other market and grid components (e.g. loads, buses, branches, grid units, market participants, etc.). Therefore, our motivation in this dissertation is to utilize the flexibility of MAS with more agents to monitor and control individual types of the modeled market and grid components. Beyond the above MAS research work for the DGs' power dispatch control, where mainly the DG agent is employed, there are several recent research works focusing on extensions of agent types as well as multi-level interactions for distributed power systems, which is consistent with our motivation. For the transition control of microgrids from the grid-connected mode to islanded mode, Cai et al. [12] presented a MAS scheme with five agent types consisting of Grid Agent, Central Agent, Generation Agent, Load Agent and Breaker Agent. Based on the five agent types, authors demonstrated a hierarchical control architecture to implement microgrid restoration. Also Dou and Liu [101] proposed a MAS-based hierarchical hybrid control framework for the power dispatch management. The framework consists of three hybrid control hierarchies, where the low-level agents control generation units, storage units and load units in a decentralized manner; the middle-level agents coordinate the operational mode of individual units in groups that build up local grid regions, thereby stabilizing secure voltages of local regions; the central upper-level agent then maximize economical benefits of the whole power system based on a multi-objective optimization. In comparison to the hierarchical architecture of Cai et al. [12] with a central agent, Ren et al. [77] proposed a decentralized control scheme that comprises five agent types for simulating five key grid components, i.e. the substation, the busbar, the feeder, the load and the generator, where no central controller is required in the MAS. Authors proposed a distributed MAS architecture, in which each agent only presents the local information of the corresponding grid component rather than the global information of the whole grid network and three agent cooperation mechanisms (i.e. connection, disconnection and power management) are used for optimizing local power dispatch decisions of individual grid components and collective decisions on the global power balance of the whole grid network.

We noticed that the most contributions of MAS-based power dispatch algorithms were made with the focus on microgrids [66, 20, 53, 75, 34, 33, 12, 13]. The main goal of this dissertation is to employ a MAS-based distributed control architecture for large power systems with distributed grid units and corresponding market units. Furthermore, the focus of our work is to discuss a distributed control architecture for not only solving the optimal power flow (OPF) problem [68, 69] or economic dispatch problems [105], but also stabilizing market prices towards a distributed market-grid coupling.

2.3 SHORT-TERM LOAD FORECASTING FOR AN ACCURATE LOAD MODELING

Short-term load forecasting (STLF) is imperative in competitive power markets [43, 31], not only for the high penetration of renewable energy [49, 14], but also for the decentralization of power systems, such as integration of microgrids [2, 41]. STLF is challenging due to the seasonality within the time series load data at different time scales and the influence of important exogenous influencing factors (e.g. weather conditions and social events) [93]. In the literature, STLF methods are classified differently. Hippert et al. [43] specified the STLF classification with two categories in terms of forecasting models: time series load models as a function of its historical values and causal load models as a function of exogenous influencing factors. According to forecasting approaches and techniques, Weron [93] summarized STLF methods into two broad categories consisting of standard statistical approaches and AI-based (artificial intelligence) techniques. Recently, more and more research work focused on hybrid approaches that combine statistical and AI-based techniques, in order to model the nonlinear relationships between power load and various influencing factors, such as PSO-based (particle swarm optimization) ARMAX model [46], support vector machine (SVM) with SOM-based (self-organized map) day type clustering [29], Kalman filter trained artificial neural networks (ANN) [38], ARIMA model with SVM-based deviation correction [70], PSO-trained SVM [85], PSO-trained ANN [4], GCP-based (grey correlation projection) random forest (RF) [97], DEMD-based (differential empirical mode decomposition) forecasting method combining support vector regression

(SVR) and auto regression (AR) [28], all these works proposed hybrid approaches for a STLF.

In addition to examining the forecasting methods, researchers also investigated the impact of exogenous influencing factors for a STLF. Already in the 90's [87, 91, 86, 52], existing literature pointed out that besides climate and seasonal variables, human activities are the primary influencing factors in load forecasting models. Also recently, Grandjean et al. [37] and Ding et al. [21, 22] illustrated that electric load curves in particular at household's level strongly feature aperiodicity caused by the uncertainty of human behaviors and activities. In general, the activity information that has been studied in the current research can be classified into two levels: 1) *high aggregation level* [71] – social activities refer to special events and day types including public holidays; 2) *appliance level* [37, 3, 6] – appliance usage behaviors refer to residents' presence and appliance TOU (time-of-use) profiles. However, these load forecasts and activity information have been focused either on a high aggregation level or an individual appliance level, and very few end-users fully understand the impact of their individual activities on power consumption and distribution [96, 30]. In this dissertation, we propose an activity-enhanced load forecasting framework that is complementary to the state-of-the-art activity-related bottom-up methods and concerns itself with the correlation between power load and human activity at different aggregation levels, in order to improve the STLF performance.

As mentioned before, at the high aggregation level, most STLF research work interpreted human (social) activities as different day types, such as weekdays, weekends and public holidays, etc. For instance, Al-Shareef et al. [1] took special religious activities as features into account for the STLF; Hong et al. [44] discussed the load seasonality affected by the seasonality of human activities in terms of month type, day type and hour type; For the load flow analysis in distribution networks, Villafafila-Robles et al. [90] suggested considering human activities as a time-dependent factor (such as seasonality) for e.g. the seasonal segmentation within a load forecasting model; Also a recent review written by Raza and Khosravi [76] stated that the human activity information in the most STLF research work is mainly encapsulated in the variable of day type and hour type. In order to explore influences of the activity information on the load forecasting, we propose to incorporate human activities on different scales directly into STLF, rather than utilize a seasonal or temporal factor instead. Most of the current research studies for understanding the causal influence of human behaviors and activities on domestic electricity demand refer to the appliance-level power consumption, such as conversion schemes from activities to appliance-level power demand [94], occupant trace based load disaggregation [54] and JMeasure-based associations mining between appliance-level power consumption and human usage behaviors, and other causal factors [82]. One research work of Paatero and Lund [72] that relates to our proposal, presented a bottom-up load modeling approach and focused on the power load in individual households. For analyzing human activities, authors defined a hourly probability factor and a social random factor for modeling activity levels of individual appliances and communal behaviors of individual households, respectively. Analogously to the work of Paatero and Lund, our proposed activity-enhanced STLF framework builds on previous work on bottom-up load modeling, in particular TOU-based load models.

Dong et al. [23] have successfully applied support vector machines (SVM) for forecasting the monthly power consumption of commercial buildings. This idea has been extended to support vector regression (SVR) and least-squares support vector machines (LS-SVM) in the work of Edwards et al. [25] to perform load forecasting of residential buildings. The forecasting was conducted based on the Campbell Creek data set which contained 15min-level consumption data of three residential homes. Remarkably, the data set comprises whole building consumption and the data of 140 other sensors for TOU (such as appliance-level consumption) and human activity (such as cupboard door-opening/-closing events) data. The work shown that LS-SVM outperforms other approaches such as neural networks.

Richardson et al. [78] have investigated a general model of residential energy consumption by using appliance-level consumption data and nationwide statistics. The authors developed a model of residential power use based on active occupancy, appliance-level TOU data and daily activity profiles incl. cooking, watching TV, laundry, cleaning, ironing, washing, etc. They have shown good performance of their model for aggregate load forecasting of a whole building with 22 houses based on a one-year data set at 1-minute intervals.

Saldanha and Beausoleil-Morrison [84] have studied individual consumption patterns in residential homes based on an 1min-level data set of the total consumption of a building and appliance-level TOU data of 12 Canadian homes. They found out that patterns between houses range widely on an annual base, while the temporal distribution of individual house's load varies significantly on a daily basis.

Blight and Coley [10] demonstrated the correlation between human behavior and domestic heating based on a data set of TOU (heating consumption, appliance-level consumption and lighting consumption) and human activity (room-level occupancy and door-opening/-closing events) data from 100 passive house units.

Another related work on modeling occupant's energy-related behaviors in residential homes is presented by Wilke et al. [95]. Using Markov-process modeling of human activities (such as work, school, cooking, housework, gardening, childcare), the authors demonstrated the predictability of time-dependent human activities based on data from a TOU survey. Furthermore, they devised a calibration methodology to improve the model's predictive power by iteratively fitting their model to data of sub-populations of their TOU survey. Also Muratori et al. [64] presented a heterogeneous Markov chain for modeling 9 types of human activities and predicting the corresponding power consumption based on activity to power conversion factors [94], as well as demonstrated an in-sample validation based on a TOU data set collected by the U.S. Bureau of Labor Statistics.

In a recent research contribution [45], Horn et al. took human consumer behaviors as appliance-level usage activities into a regression-based load forecasting model of individual appliances, where the activity information in terms of working modes is inferred based on the smart meter data of individual appliances collected in the CoSSMic project.

2.4 SUMMARY

In this chapter, we presented the related work and literature in the research field of ODD, MAS and STLF. Each field provides an important component, in terms of the methods and technologies for the development of our proposed market-grid coupling. To our knowledge, previous research does not confirm any of the defined theses regarding the market-grid coupling, as the coupling aspect between the market and the grid has not been investigated yet. However, the current research outcomes of all three fields offer methodical insights regarding system components of a feedback control concept. In general, our work extends the proposed feedback control loop of the ODD research work with a market module, in order to study the impact of optimal dynamic dispatch on the market price stability as well as the importance of dynamic pricing for achieving a steady-state stability of the grid. This provides a possibility to investigate Thesis 1. Furthermore, we employ a MAS-based distributed control architecture for large power systems with distributed grid units and corresponding market units, which distinguishes our work from the current ODD research works with a distributed MPC strategy (Thesis 2). Finally, we proposed an activity-enhanced load forecasting framework that is complementary to the state-of-the-art activity-related bottom-up methods of the load modeling, in order to improve the STLF performance. This proposed load forecasting framework provides a well-defined integrated process for the enhancement of the MPC-based power dispatch control (Thesis 3).

Bibliography

[1] AJ Al-Shareef, EA Mohamed, and E Al-Judaibi. One hour ahead load forecasting using artificial neural network for the western area of saudi arabia. *International Journal of Electrical Systems Science and Engineering*, 1(1):35–40, 2008.

[2] N. Amjady, F. Keynia, and H. Zareipour. Short-term load forecast of microgrids by a new bilevel prediction strategy. *Smart Grid, IEEE Transactions on*, 1(3):286–294, Dec 2010. ISSN 1949-3053. doi: 10.1109/TSG.2010.2078842.

[3] Nicoleta Arghira, Lamis Hawarah, St?phane Ploix, and Mireille Jacomino. Prediction of appliances energy use in smart homes. *Energy*, 48(1):128 – 134, 2012. ISSN 0360-5442. 6th Dubrovnik Conference on Sustainable Development of Energy Water and Environmental Systems, {SDEWES} 2011.

[4] Ellen Banda and Komla A. Folly. Short term load forecasting based on hybrid ann and pso. In Ying Tan, Yuhui Shi, Fernando Buarque, Alexander Gelbukh, Swagatam Das, and Andries Engelbrecht, editors, *Advances in Swarm and Computational Intelligence*, volume 9142 of *Lecture Notes in Computer Science*, pages 98–106. Springer International Publishing, 2015. ISBN 978-3-319-20468-0. doi: 10.1007/978-3-319-20469-7_12. URL http://dx.doi.org/10.1007/978-3-319-20469-7_12.

[5] M.E. Baran and I.M. El-Markabi. A multiagent-based dispatching scheme for distributed generators for voltage support on distribution feeders. *Power Systems, IEEE Transactions on*, 22(1):52–59, Feb 2007. ISSN 0885-8950. doi: 10.1109/TPWRS.2006.889140.

[6] M. Basu. Hybridization of bee colony optimization and sequential quadratic programming for dynamic economic dispatch. *International Journal of Electrical Power & Energy Systems*, 44(1):591 – 596, 2013. ISSN 0142-0615. doi: http://dx.doi.org/10.1016/j.ijepes.2012.08.026.

[7] T.E. Bechert and H.G. Kwatny. On the optimal dynamic dispatch of real power. *Power Apparatus and Systems, IEEE Transactions on*, PAS-91(3):889–898, May 1972. doi: 10.1109/TPAS.1972.293422.

[8] A. Bidram, A. Davoudi, F.L. Lewis, and Zhihua Qu. Secondary control of microgrids based on distributed cooperative control of multi-agent systems.

Generation, Transmission Distribution, IET, 7(8):822–831, Aug 2013. ISSN 1751-8687. doi: 10.1049/iet-gtd.2012.0576.

[9] B. Biegel, J. Stoustrup, J. Bendtsen, and P. Andersen. Model predictive control for power flows in networks with limited capacity. In *American Control Conference (ACC), 2012*, pages 2959–2964, June 2012. doi: 10.1109/ACC.2012. 6314854.

[10] Thomas S. Blight and David A. Coley. Sensitivity analysis of the effect of occupant behaviour on the energy consumption of passive house dwellings. *Energy and Buildings*, 66(0):183 – 192, 2013. ISSN 0378-7788.

[11] A. Botterud, M.R. Mahalik, T.D. Veselka, Heon-Su Ryu, and Ki-Won Sohn. Multi-agent simulation of generation expansion in electricity markets. In *Power Engineering Society General Meeting, 2007. IEEE*, pages 1–8, June 2007. doi: 10.1109/PES.2007.385566.

[12] Niannian Cai, Xufeng Xu, and J. Mitra. A hierarchical multi-agent control scheme for a black start-capable microgrid. In *Power and Energy Society General Meeting, 2011 IEEE*, pages 1–7, July 2011. doi: 10.1109/PES.2011.6039570.

[13] Niannian Cai, Nguyen Thi Thanh Nga, and J. Mitra. Economic dispatch in microgrids using multi-agent system. In *North American Power Symposium (NAPS), 2012*, pages 1–5, Sept 2012. doi: 10.1109/NAPS.2012.6336435.

[14] J.P.S. Catalao, H.M.I. Pousinho, and V.M.F. Mendes. Short-term wind power forecasting in portugal by neural networks and wavelet transform. *Renewable Energy*, 36(4):1245 – 1251, 2011. ISSN 0960-1481. doi: http://dx.doi.org/ 10.1016/j.renene.2010.09.016. URL http://www.sciencedirect.com/ science/article/pii/S0960148110004477.

[15] C.M. Colson and M.H. Nehrir. Algorithms for distributed decision-making for multi-agent microgrid power management. In *Power and Energy Society General Meeting, 2011 IEEE*, pages 1–8, July 2011. doi: 10.1109/PES.2011. 6039764.

[16] C.M. Colson and M.H. Nehrir. Comprehensive real-time microgrid power management and control with distributed agents. *Smart Grid, IEEE Transactions on*, 4(1):617–627, March 2013. ISSN 1949-3053. doi: 10.1109/TSG.2012. 2236368.

[17] G. Conzelmann, G. Boyd, V. Koritarov, and T. Veselka. Multi-agent power market simulation using emcas. In *Power Engineering Society General Meeting, 2005. IEEE*, pages 2829–2834 Vol. 3, June 2005. doi: 10.1109/PES.2005. 1489271.

[18] Alejandro J. del Real, Alicia Arce, and Carlos Bordons. Combined environmental and economic dispatch of smart grids using distributed model predictive control. *International Journal of Electrical Power & Energy Systems*, 54(0):65 – 76, 2014. ISSN 0142-0615. doi: http://dx.doi.org/10.1016/j.ijepes.2013.06.035.

[19] Jean-Yves Dieulot, Frederic Colas, Lamine Chalal, and Genevieve Dauphin-Tanguy. Economic supervisory predictive control of a hybrid power generation plant. *Electric Power Systems Research*, 127:221 – 229, 2015. ISSN 0378-7796. doi: http://dx.doi.org/10.1016/j.epsr.2015.06.006. URL http://www.sciencedirect.com/science/article/pii/S0378779615001819.

[20] Aris L. Dimeas and Nikos D. Hatziargyriou. Operation of a multiagent system for microgrid control. *IEEE TRANSACTIONS ON POWER SYSTEMS*, 20:1447–1455, August 2005.

[21] Yong Ding, Martin Alexander Neumann, Per Goncalves Da Silva, and Michael Beigl. A framework for short-term activity-aware load forecasting. In *Joint Proceedings of the Workshop on AI Problems and Approaches for Intelligent Environments and Workshop on Semantic Cities*, AIIP '13, pages 23–28, New York, NY, USA, 2013. ACM.

[22] Yong Ding, Martin A Neumann, Ömer Kehri, Geoff Ryder, Till Riedel, and Michael Beigl. From load forecasting to demand response-a web of things use case. In *Proceedings of the 5th International Workshop on Web of Things*, pages 28–33. ACM, 2014.

[23] Bing Dong, Cheng Cao, and Siew Eang Lee. Applying support vector machines to predict building energy consumption in tropical region. *Energy and Buildings*, 37(5):545–553, 2005. ISSN 0378-7788. doi: 10.1016/j.enbuild.2004.09.009.

[24] Chunxia Dou, Chunchun Mao, Zhiqian Bo, and Xingzhong Zhang. A multi-agent model based decentralized coordinated control for large power system transient stability improvement. In *Universities Power Engineering Conference (UPEC), 2010 45th International*, pages 1–5, Aug 2010.

[25] Richard E. Edwards, Joshua New, and Lynne E. Parker. Predicting future hourly residential electrical consumption: A machine learning case study. *Energy and Buildings*, 49(0):591 – 603, 2012. ISSN 0378-7788.

[26] A.M. Elaiw, X. Xia, and A.M. Shehata. Application of model predictive control to optimal dynamic dispatch of generation with emission limitations.

Electric Power Systems Research, 84(1):31 – 44, 2012. ISSN 0378-7796. doi: http://dx.doi.org/10.1016/j.epsr.2011.09.024.

[27] A.M. Elaiw, A.M. Shehata, and M.A. Alghamdi. A model predictive control approach to combined heat and power dynamic economic dispatch problem. *Arabian Journal for Science and Engineering*, 39(10):7117–7125, 2014. ISSN 1319-8025. doi: 10.1007/s13369-014-1218-0.

[28] Guo-Feng Fan, Li-Ling Peng, Wei-Chiang Hong, and Fan Sun. Electric load forecasting by the {SVR} model with differential empirical mode decomposition and auto regression. *Neurocomputing*, 173, Part 3:958 – 970, 2016. ISSN 0925-2312. doi: http://dx.doi.org/10.1016/j.neucom.2015.08.051. URL http://www.sciencedirect.com/science/article/pii/S0925231215012230.

[29] Shu Fan and Luonan Chen. Short-term load forecasting based on an adaptive hybrid method. *Power Systems, IEEE Transactions on*, 21(1):392–401, Feb 2006. ISSN 0885-8950. doi: 10.1109/TPWRS.2005.860944.

[30] Ahmad Faruqui, Sanem Sergici, and Ahmed Sharif. The impact of informational feedback on energy consumption – survey of the experimental evidence. *Energy*, 35(4):1598 – 1608, 2010. ISSN 0360-5442.

[31] EugeneA. Feinberg and Dora Genethliou. Load forecasting. In JoeH. Chow, FelixF. Wu, and James Momoh, editors, *Applied Mathematics for Restructured Electric Power Systems*, Power Electronics and Power Systems, pages 269–285. Springer US, 2005. ISBN 978-0-387-23470-0. doi: 10.1007/0-387-23471-3_12. URL http://dx.doi.org/10.1007/0-387-23471-3_12.

[32] Jacques Ferber. *Multi-agent systems: An introduction to distributed artificial intelligence*. Addison-Wesley Professional, 1999.

[33] Hassan Feroze. Multi-agent systems in microgrids: Design & implementation. Master's thesis, Virginia Tech, August 2009.

[34] T. Funabashi, T. Tanabe, T. Nagata, and R. Yokoyama. An autonomous agent for reliable operation of power market and systems including microgrids. In *Electric Utility Deregulation and Restructuring and Power Technologies, 2008. DRPT 2008. Third International Conference on*, pages 173 –177, april 2008.

[35] Matthias D. Galus and G. Andersson. Demand management of grid connected plug-in hybrid electric vehicles (phev). In *Energy 2030 Conference, 2008. ENERGY 2008. IEEE*, pages 1–8, Nov 2008. doi: 10.1109/ENERGY.2008.4781014.

[36] F. Garcia and C. Bordons. Optimal economic dispatch for renewable energy microgrids with hybrid storage using model predictive control. In *Industrial Electronics Society, IECON 2013 - 39th Annual Conference of the IEEE*, pages 7932–7937, Nov 2013. doi: 10.1109/IECON.2013.6700458.

[37] A. Grandjean, J. Adnot, and G. Binet. A review and an analysis of the residential electric load curve models. *Renewable and Sustainable Energy Reviews*, 16(9):6539 – 6565, 2012. ISSN 1364-0321.

[38] Che Guan, P.B. Luh, L.D. Michel, M.A. Coolbeth, and P.B. Friedland. Hybrid kalman algorithms for very short-term load forecasting and confidence interval estimation. In *Power and Energy Society General Meeting, 2010 IEEE*, pages 1–8, 2010.

[39] Rasmus Halvgaard, Lieven Vandenberghe, Niels K Poulsen, Henrik Madsen, and John B Jørgensen. Distributed model predictive control for smart energy systems. *Power*, 1:2, 2014.

[40] H.M. Hassan and Y.A.I. Mohamed. Market-oriented energy management of a hybrid power system via model predictive control with constraints optimizer. In *PES General Meeting — Conference Exposition, 2014 IEEE*, pages 1–5, July 2014. doi: 10.1109/PESGM.2014.6939571.

[41] Luis Hernandez, Carlos Baladrón, Javier M Aguiar, Belén Carro, Antonio J Sanchez-Esguevillas, and Jaime Lloret. Short-term load forecasting for microgrids based on artificial neural networks. *Energies*, 6(3):1385–1408, 2013.

[42] Paul Hines, Dong Jia, and Sarosh Talukdar. Distributed model predictive control for electric grids. In *Proc. of the Carnegie Mellon Transmission Conf*, 2004.

[43] H.S. Hippert, C.E. Pedreira, and R.C. Souza. Neural networks for short-term load forecasting: a review and evaluation. *Power Systems, IEEE Transactions on*, 16(1):44 –55, feb 2001.

[44] Tao Hong, Min Gui, M.E. Baran, and H.L. Willis. Modeling and forecasting hourly electric load by multiple linear regression with interactions. In *Power and Energy Society General Meeting, 2010 IEEE*, pages 1–8, July 2010. doi: 10.1109/PES.2010.5589959.

[45] Geir Horn, Salvatore Venticinque, and Alba Amato. Inferring appliance load profiles from measurements. In Giuseppe Di Fatta, Giancarlo Fortino, Wenfeng Li, Mukaddim Pathan, Frederic Stahl, and Antonio Guerrieri, editors, *Internet and Distributed Computing Systems*, volume 9258 of *Lecture*

Notes in Computer Science, pages 118–130. Springer International Publishing, 2015. ISBN 978-3-319-23236-2. doi: 10.1007/978-3-319-23237-9_11. URL http://dx.doi.org/10.1007/978-3-319-23237-9_11.

[46] Chao-Ming Huang, Chi-Jen Huang, and Ming-Li Wang. A particle swarm optimization to identifying the armax model for short-term load forecasting. *Power Systems, IEEE Transactions on*, 20(2):1126–1133, May 2005. ISSN 0885-8950. doi: 10.1109/TPWRS.2005.846106.

[47] Zhang Jian, Ai Qian, Jiang Chuanwen, Wang Xingang, Zheng Zhanghua, and Gu Chenghong. The application of multi agent system in microgrid coordination control. In *Sustainable Power Generation and Supply, 2009. SUPERGEN '09. International Conference on*, pages 1–6, April 2009. doi: 10.1109/SUPERGEN.2009.5348277.

[48] A. Jokic, M. Lazar, and P.P.J.vanden Bosch. Price-based optimal control of power flow in electrical energy transmission networks. In Alberto Bemporad, Antonio Bicchi, and Giorgio Buttazzo, editors, *Hybrid Systems: Computation and Control*, volume 4416 of *Lecture Notes in Computer Science*, pages 315–328. Springer Berlin Heidelberg, 2007. ISBN 978-3-540-71492-7. doi: 10.1007/978-3-540-71493-4_26.

[49] Tryggvi Jonsson, Pierre Pinson, and Henrik Madsen. On the market impact of wind energy forecasts. *Energy Economics*, 32(2):313 – 320, 2010. ISSN 0140-9883. doi: http://dx.doi.org/10.1016/j.eneco.2009.10.018. URL http://www.sciencedirect.com/science/article/pii/S0140988309002011.

[50] A. Khatamianfar, M. Khalid, A.V. Savkin, and V.G. Agelidis. Improving wind farm dispatch in the australian electricity market with battery energy storage using model predictive control. *Sustainable Energy, IEEE Transactions on*, 4(3):745–755, July 2013. ISSN 1949-3029. doi: 10.1109/TSTE.2013.2245427.

[51] Pascal Kienast. Optimal overload response in electric power systems applying model predictive control. Master's thesis, ETH Zurich, 2007.

[52] Kwang-Ho Kim, Hyoung-Sun Youn, and Yong-Cheol Kang. Short-term load forecasting for special days in anomalous load conditions using neural networks and fuzzy inference method. *Power Systems, IEEE Transactions on*, 15(2):559 –565, may 2000.

[53] J. K. Kok, C. J. Warmer, and I. G. Kamphuis. Powermatcher: multiagent control in the electricity infrastructure. In *Proceedings of the fourth interna-*

tional joint conference on Autonomous agents and multiagent systems, AAMAS '05, pages 75–82, New York, NY, USA, 2005. ACM. ISBN 1-59593-093-0.

[54] Seungwoo Lee, Daye Ahn, Sukjun Lee, Rhan Ha, and Hojung Cha. Personalized energy auditor: Estimating personal electricity usage. In *Pervasive Computing and Communications (PerCom), 2014 IEEE International Conference on*, pages 44–49, March 2014. doi: 10.1109/PerCom.2014.6813942.

[55] Jiaqi Liang, R.G. Harley, and G.K. Venayagamoorthy. Adaptive critic design based dynamic optimal power flow controller for a smart grid. In *Computational Intelligence Applications In Smart Grid (CIASG), 2011 IEEE Symposium on*, pages 1–8, April 2011. doi: 10.1109/CIASG.2011.5953340.

[56] T. Logenthiran, D. Srinivasan, and D. Wong. Multi-agent coordination for der in microgrid. In *Sustainable Energy Technologies, 2008. ICSET 2008. IEEE International Conference on*, pages 77–82, Nov 2008. doi: 10.1109/ICSET.2008. 4746976.

[57] T. Logenthiran, D. Srinivasan, A.M. Khambadkone, and H.N. Aung. Scalable multi-agent system (mas) for operation of a microgrid in islanded mode. In *Power Electronics, Drives and Energy Systems (PEDES) 2010 Power India, 2010 Joint International Conference on*, pages 1–6, Dec 2010. doi: 10.1109/PEDES.2010.5712459.

[58] T. Logenthiran, Dipti Srinivasan, and Ashwin M. Khambadkone. Multi-agent system for energy resource scheduling of integrated microgrids in a distributed system. *Electric Power Systems Research*, 81(1):138 – 148, 2011. ISSN 0378-7796. doi: http://dx.doi.org/10.1016/j.epsr.2010.07. 019. URL http://www.sciencedirect.com/science/article/pii/ S0378779610001823.

[59] M.A. Lopez, S. Martin, J.A. Aguado, and S. de la Torre. Market-oriented operation in microgrids using multi-agent systems. In *Power Engineering, Energy and Electrical Drives (POWERENG), 2011 International Conference on*, pages 1–6, May 2011. doi: 10.1109/PowerEng.2011.6036418.

[60] S.D.J. McArthur, E.M. Davidson, V.M. Catterson, A.L. Dimeas, N.D. Hatziargyriou, F. Ponci, and T. Funabashi. Multi-agent systems for power engineering applications – part i: Concepts, approaches, and technical challenges. *Power Systems, IEEE Transactions on*, 22(4):1743 –1752, nov. 2007.

[61] Zhou Ming, Ren Jianwen, Li Gengyin, and Xu Xianghai. A multi-agent based dispatching operation instructing system in electric power systems.

In *Power Engineering Society General Meeting, 2003, IEEE*, volume 1, pages 436–440 Vol. 1, July 2003. doi: 10.1109/PES.2003.1267215.

[62] D. Mst, W. Fichtner, M. Ragwitz, and D. Veit, editors. *New methods for energy market modelling : Proceedings of the First European Workshop on Energy Market Modelling using Agent-Based Computational Economics*, 2007.

[63] R. Mudumbai, S. Dasgupta, and B.B. Cho. Distributed control for optimal economic dispatch of a network of heterogeneous power generators. *Power Systems, IEEE Transactions on*, 27(4):1750–1760, Nov 2012. doi: 10.1109/TPWRS.2012.2188048.

[64] Matteo Muratori, Matthew C. Roberts, Ramteen Sioshansi, Vincenzo Marano, and Giorgio Rizzoni. A highly resolved modeling technique to simulate residential power demand. *Applied Energy*, 107(0):465 – 473, 2013. ISSN 0306-2619.

[65] T. Nagata and H. Sasaki. A multi-agent approach to power system restoration. *Power Systems, IEEE Transactions on*, 17(2):457–462, May 2002. ISSN 0885-8950. doi: 10.1109/TPWRS.2002.1007918.

[66] T. Nagata, Y. Tao, H. Sasaki, and H. Fujita. A multi-agent approach to distribution system restoration. *Power Engineering Society General Meeting*, 2: 660, 2003.

[67] T. Nagata, Y. Tao, K. Kimura, H. Sasaki, and H. Fujita. A multi-agent approach to distribution system restoration. In *Circuits and Systems, 2004. MWSCAS '04. The 2004 47th Midwest Symposium on*, volume 2, pages II–333–II–336 vol.2, July 2004. doi: 10.1109/MWSCAS.2004.1354161.

[68] P.H. Nguyen, W.L. Kling, G. Georgiadis, M. Papatriantafilou, L.A. Tuan, and L. Bertling. Distributed routing algorithms to manage power flow in agent-based active distribution network. In *Innovative Smart Grid Technologies Conference Europe (ISGT Europe), 2010 IEEE PES*, pages 1–7, Oct 2010. doi: 10.1109/ISGTEUROPE.2010.5638951.

[69] P.H. Nguyen, W.L. Kling, and P.F. Ribeiro. Smart power router: A flexible agent-based converter interface in active distribution networks. *Smart Grid, IEEE Transactions on*, 2(3):487–495, Sept 2011. ISSN 1949-3053. doi: 10.1109/ TSG.2011.2159405.

[70] Hongzhan Nie, Guohui Liu, Xiaoman Liu, and Yong Wang. Hybrid of {ARIMA} and {SVMs} for short-term load forecasting. *Energy Procedia*, 16, Part C:1455 – 1460, 2012. ISSN 1876-6102. doi: http://dx.doi.org/10.1016/j.

egypro.2012.01.229. URL `http://www.sciencedirect.com/science/article/pii/S1876610212002391`. 2012 International Conference on Future Energy, Environment, and Materials.

[71] Dongxiao Niu, Yongli Wang, and Desheng Dash Wu. Power load forecasting using support vector machine and ant colony optimization. *Expert Systems with Applications*, 37(3):2531 – 2539, 2010. ISSN 0957-4174.

[72] Jukka V. Paatero and Peter D. Lund. A model for generating household electricity load profiles. *International Journal of Energy Research*, 30(5):273–290, 2006. ISSN 1099-114X. doi: 10.1002/er.1136. URL `http://dx.doi.org/10.1002/er.1136`.

[73] A. Pantoja and N. Quijano. A population dynamics approach for the dispatch of distributed generators. *Industrial Electronics, IEEE Transactions on*, 58(10):4559–4567, Oct 2011. ISSN 0278-0046. doi: 10.1109/TIE.2011.2107714.

[74] Panagiotis Patrinos, Sergio Trimboli, and Alberto Bemporad. Stochastic mpc for real-time market-based optimal power dispatch. In *Decision and Control and European Control Conference (CDC-ECC), 2011 50th IEEE Conference on*, pages 7111–7116, Dec 2011. doi: 10.1109/CDC.2011.6160798.

[75] Zhifeng Qiu, Geert Deconinck, Ning Gui, and Ronnie Belmans. A multi-agent system architecture for electrical energy matching in a microgrid. in Proceedings of Fourth IEEE Young Researchers Symposium in Electrical Power Engineering, 2008.

[76] Muhammad Qamar Raza and Abbas Khosravi. A review on artificial intelligence based load demand forecasting techniques for smart grid and buildings. *Renewable and Sustainable Energy Reviews*, 50:1352 – 1372, 2015. ISSN 1364-0321. doi: http://dx.doi.org/10.1016/j.rser.2015.04.065. URL `http://www.sciencedirect.com/science/article/pii/S1364032115003354`.

[77] Fenghui Ren, Minjie Zhang, and D. Sutanto. A multi-agent solution to distribution system management by considering distributed generators. *Power Systems, IEEE Transactions on*, 28(2):1442–1451, May 2013. ISSN 0885-8950. doi: 10.1109/TPWRS.2012.2223490.

[78] Ian Richardson, Murray Thomson, David Infield, and Conor Clifford. Domestic electricity use: A high-resolution energy demand model. *Energy and Buildings*, 42(10):1878 – 1887, 2010. ISSN 0378-7788.

[79] S. Rivera, A.M. Farid, and K. Youcef-Toumi. A multi-agent system transient stability platform for resilient self-healing operation of multiple microgrids. In *Innovative Smart Grid Technologies Conference (ISGT), 2014 IEEE PES*, pages 1–5, Feb 2014. doi: 10.1109/ISGT.2014.6816377.

[80] L. Robitzky, S.C. Muller, S. Dalhues, U. Hager, and C. Rehtanz. Agent-based redispatch for real-time overload relief in electrical transmission systems. In *Power Energy Society General Meeting, 2015 IEEE*, pages 1–5, July 2015. doi: 10.1109/PESGM.2015.7285886.

[81] R. Roche, B. Blunier, A. Miraoui, V. Hilaire, and A. Koukam. Multi-agent systems for grid energy management: A short review. In *IECON 2010 - 36th Annual Conference on IEEE Industrial Electronics Society*, pages 3341 –3346, nov. 2010.

[82] S. Rollins and N. Banerjee. Using rule mining to understand appliance energy consumption patterns. In *Pervasive Computing and Communications (PerCom), 2014 IEEE International Conference on*, pages 29–37, March 2014. doi: 10.1109/PerCom.2014.6813940.

[83] J. Sachs, M. Sonntag, and O. Sawodny. Two layer model predictive control for a cost efficient operation of island energy systems. In *American Control Conference (ACC), 2015*, pages 4941–4946, July 2015. doi: 10.1109/ACC.2015. 7172108.

[84] Neil Saldanha and Ian Beausoleil-Morrison. Measured end-use electric load profiles for 12 canadian houses at high temporal resolution. *Energy and Buildings*, 49(0):519 – 530, 2012. ISSN 0378-7788.

[85] A. Selakov, D. Cvijetinovic, L. Milovic, S. Mellon, and D. Bekut. Hybrid pso-svm method for short-term load forecasting during periods with significant temperature variations in city of burbank. *Applied Soft Computing*, 16:80 – 88, 2014. ISSN 1568-4946. doi: http://dx.doi.org/10.1016/j.asoc.2013.12. 001. URL http://www.sciencedirect.com/science/article/pii/ S1568494613004183.

[86] D. Srinivasan, Swee Sien Tan, C.S. Cheng, and Eng Kiat Chan. Parallel neural network-fuzzy expert system strategy for short-term load forecasting: system implementation and performance evaluation. *Power Systems, IEEE Transactions on*, 14(3):1100 –1106, aug 1999.

[87] Dipti Srinivasan, C.S. Chang, and A.C. Liew. Demand forecasting using fuzzy neural computation, with special emphasis on weekend and public

holiday forecasting. *Power Systems, IEEE Transactions on*, 10(4):1897 –1903, nov. 1995.

[88] Kenneth Van den Bergh, Dries Couckuyt, Erik Delarue, and William D'haeseleer. Redispatching in an interconnected electricity system with high renewables penetration. *Electric Power Systems Research*, 2015.

[89] A.N. Venkat, I.A. Hiskens, J.B. Rawlings, and S.J. Wright. Distributed mpc strategies with application to power system automatic generation control. *Control Systems Technology, IEEE Transactions on*, 16(6):1192–1206, Nov 2008. doi: 10.1109/TCST.2008.919414.

[90] R. Villafafila-Robles, A. Sumper, B. Bak-Jensen, and E. Valsera-Naranjo. Probabilistic analysis in normal operation of distribution system with distributed generation. In *Power and Energy Society General Meeting, 2011 IEEE*, pages 1–8, July 2011. doi: 10.1109/PES.2011.6039654.

[91] A.J Wang and B Ramsay. A neural network based estimator for electricity spot-pricing with particular reference to weekend and public holidays. *Neurocomputing*, 23(1):47 – 57, 1998.

[92] Kaiyan Wang, Xianjue Luo, Ling Wu, and Xuchen Liu. Optimal dispatch of wind-hydro-thermal power system with priority given to clean energy. *Proceedings of the CSEE*, 13:006, 2013.

[93] R. Weron. *Modeling and forecasting electricity loads and prices: A statistical approach*, volume 403 of *The Wiley Finance Series*. John Wiley & Sons, 2006.

[94] Joakim Widn and Ewa Wckelgard. A high-resolution stochastic model of domestic activity patterns and electricity demand. *Applied Energy*, 87(6): 1880 – 1892, 2010. ISSN 0306-2619.

[95] Urs Wilke, Frdric Haldi, Jean-Louis Scartezzini, and Darren Robinson. A bottom-up stochastic model to predict building occupants' time-dependent activities. *Building and Environment*, 60(0):254 – 264, 2013. ISSN 0360-1323.

[96] G. Wood and M. Newborough. Dynamic energy-consumption indicators for domestic appliances: environment, behaviour and design. *Energy and Buildings*, 35(8):821 – 841, 2003. ISSN 0378-7788. doi: 10.1016/S0378-7788(02) 00241-4. URL http://www.sciencedirect.com/science/article/pii/S0378778802002414.

[97] Xiaoyu Wu, Jinghan He, T. Yip, and Pei Zhang. A two-stage random forest method for short-term load forecasting. In *PowerTech, 2015 IEEE Eindhoven*, pages 1–6, June 2015. doi: 10.1109/PTC.2015.7232530.

[98] X. Xia and A.M. Elaiw. Optimal dynamic economic dispatch of generation: A review. *Electric Power Systems Research*, 80(8):975 – 986, 2010. ISSN 0378-7796. doi: http://dx.doi.org/10.1016/j.epsr.2009.12.012.

[99] Xiaohua Xia, Jiangfeng Zhang, and A. Elaiw. A model predictive control approach to dynamic economic dispatch problem. In *PowerTech, 2009 IEEE Bucharest*, pages 1–7, June 2009. doi: 10.1109/PTC.2009.5282270.

[100] Xiaohua Xia, Jiangfeng Zhang, and Ahmed Elaiw. An application of model predictive control to the dynamic economic dispatch of power generation. *Control Engineering Practice*, 19(6):638 – 648, 2011. ISSN 0967-0661. doi: http://dx.doi.org/10.1016/j.conengprac.2011.03.001. {SAFEPROCESS} 2009 Special Section: Fault Diagnosis Systems (7th {IFAC} Symposium on Fault Detection, Supervision and Safety of Technical Processes (SAFEPROCESS) in Barcelona, 30th June to 3rd July 2009).

[101] Chun xia Dou and Bin Liu. Multi-agent based hierarchical hybrid control for smart microgrid. *Smart Grid, IEEE Transactions on*, 4(2):771–778, June 2013. ISSN 1949-3053. doi: 10.1109/TSG.2012.2230197.

[102] Le Xie and M.D. Ilic. Model predictive economic/environmental dispatch of power systems with intermittent resources. In *Power Energy Society General Meeting, 2009. PES '09. IEEE*, pages 1–6, July 2009. doi: 10.1109/PES.2009.5275940.

[103] Nanpeng Yu and Chen-Ching Liu. Multi-agent systems and electricity markets: State-of-the-art and the future. In *Power and Energy Society General Meeting - Conversion and Delivery of Electrical Energy in the 21st Century, 2008 IEEE*, pages 1–2, July 2008. doi: 10.1109/PES.2008.4596296.

[104] Tie-jiang YUAN, Qin CHAO, Yibulayin TUERXUN, and Yi-yan LI. Electricity market-oriented optimization of environmental economic dispatching for power grid containing wind power [j]. *Power System Technology*, 20:025, 2009.

[105] Ziang Zhang, Xichun Ying, and Mo yuen Chow. Decentralizing the economic dispatch problem using a two-level incremental cost consensus algorithm in a smart grid environment. In *North American Power Symposium (NAPS), 2011*, pages 1–7, Aug 2011. doi: 10.1109/NAPS.2011.6025103.

3

A Market-Grid Coupling
Definition and Requirements

3.1 ABSTRACT AND CONTEXT

This chapter provides a comprehensive description of the power market modeling and the power grid modeling employed in this dissertation. The modeling approaches of both market dynamics and grid dynamics enable a price-based feedback system towards an economic power dispatch, which are considered as the first requirements for the market-grid coupling. In order to realize an interoperable control between the power market and the power grid, we extend the economic power dispatch problem with a market price control task. Based on the model description of both the market and grid as well as their relationship in both models, a formal definition of the market-grid coupling is presented. Additionally, we outline the technical requirements of a feedback control system for implementing the defined market-grid coupling concept. The content of this chapter is based on the VDAR[1] project results, a paper published at IEEE SmartGridComm 2013 [14] and a paper published at IEEE ENERGYCON 2016 [15].

---

1 http://www.teco.edu/research/vdar/, funded by BMBF (Federal Ministry of Education and Research)

3.2 THE POWER MARKET

The power market in this dissertation refers to the trading platform for the electricity, which focuses mainly on the economic aspects of the power exchange and where contracts are made between sellers and buyers for the power delivery. In Germany, the Leipzig Energy Exchange (European Energy Exchange - EEX) is the main arena for trading power with a Double Auction (DA) pricing mechanism [39]. Recently, the EEX has become an international trading platform with more than 200 trading participants from about 30 countries. At the EEX, electricity can be traded either on the *derivatives* market or on the *spot* market depending on the power delivery period. On the derivatives market, long-term power delivery contracts can be made with a lead time of up to 6 years in the future, while the spot market is used for trading the short-term deliverable power within 1-2 days in terms of day-ahead trading and intraday trading. As this dissertation is concerned with the coupling between the economic power market and the physical power grid with real-time operations, we are interested only in the spot market due to its short-term response.

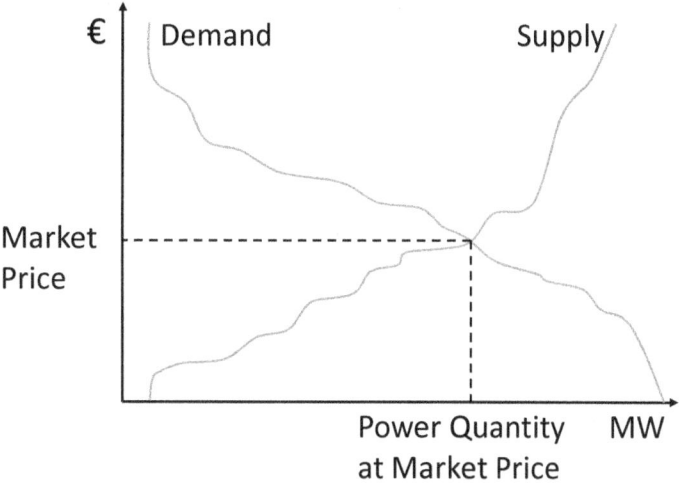

Figure 3.: Principle of market price calculation through matching supply and demand curves

The spot market consists of a day-ahead market and a intraday market. The day-ahead market is used as the trading platform of the power delivery for the next day. The day-ahead market operates based on a blind auction process, with which aggregate supply and demand curves will be matched once a day and thus market prices will be determined in an anonymous and transparent manner. Figure 3 illustrates this price matching. The market clearing price is calculated based on the merit order principle for each hour or each block[2] of the next day. Market participants submit their orders (auction bids) for the hourly auction of power quantities into the order book. In Germany, the deadline for submitting orders is 12:00 CET. Then, the day-ahead market calculates the offer and demand curves and their intersection for each hour of the next day, afterwards publishing results at 12:55 CET every day. From 15:00 CET on, it is already possible to operate intraday trading for the next day.

The intraday market is used for trading the power delivery on the day of operation or the next day. In comparison with the trading mechanism of the day-ahead market, which is based on the principle of the market clearing price, the intraday market is organized by continuous trading that determines market prices based on the latest placed orders for all transactions. Therefore, market prices in the intraday trading are determined by the "pay as bid" method and termed also as bid prices. In Germany, the intraday trading starts at 15:00 CET of the previous day with the continuous hourly intraday trading as well as the 15-minute intraday call auction that will be continuously traded first from 16:00 CET as the continuous quarterly intraday trading. In addition to hour and 15-minute contracts, block contracts are possible for the trading. All contracts can be traded up to 30 minutes before it is due for the physical power delivery. The intraday market is particularly relevant as the grid penetration of intermittent renewables grows. In order to discover short-term and unpredictable changes in the power generation and demand, the intraday trading will be used as a market-based mechanism, before the use of balancing power is required.

When unexpected deviations between the scheduled power production and consumption result in system imbalances that the intraday market cannot cover, the TSOs (Transmission System Operators) have to use emergency reserves (also known as balancing power) from the balancing power market to restore the supply and demand balance. In Germany, these reserves balance the fluctuations of the power grid in three different frequency control stages [37, 25], i.e. *primary control* with a response time within 30 s, *secondary control* with a response time within 5 min and *tertiary control* (minute reserve) with a manual response time within 15 min. The timing of the reserves for the frequency control is depicted in Figure 4. As shown in the figure, in Germany, the TSOs are responsible for the provision of reserves in terms of primary, secondary and tertiary control, only within the first 4 quarter hours after occurrence of a power imbalance. After 60 minutes, the BRPs

---

2 The definition of blocks from EPEX SPOT: "block orders are used to link several hours on an all-or-none basis, which means that either the bid is matched on all of the hours or it is entirely rejected. ... A block order is executed for its full quantity only. ..."

(Balance Responsible Parties) [25] take full responsibility for the compensation of a power imbalance.

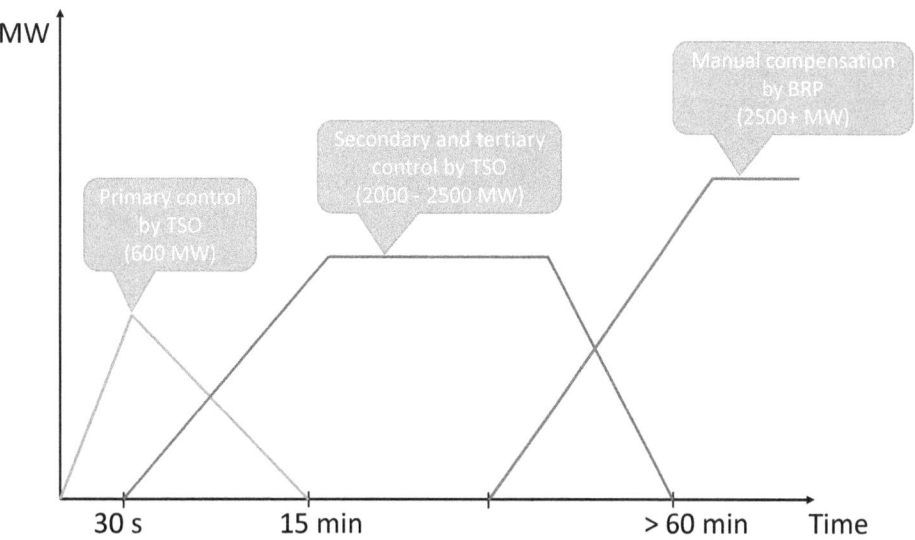

Figure 4.: Scheme of the timing and capacity of the different reserve types in Germany

As we notice that the intraday and balancing markets can deal with a grid relief to a certain extent, in order to prevent the physical grid from a collapse (e.g. a blackout). Due to the demand uncertainty (i.e. load volatility) and the uncertainty of forecasts for renewable feed-ins, these short-term power markets need to be more aware of the real-time information of the physical grid in terms of generation capacity and transmission constraints, and to be more flexible to effectively respond to increased grid uncertainties [9]. Therefore, Borggrefe and Neuhoff [9] suggested to integrate the demand side into both markets and to enhance the market interaction with physical grid components on a intraday basis for an intraday optimization of the real-time power dispatch. For this purpose, there are several proposals towards a new market design for the European power market. Two important approaches with on-going projects that are detailed in the following, consider more market-grid interactions, which can also confirm the relevance of the objective in this dissertation.

One of the approaches is *market coupling* [16] that creates technical conditions for optimizing the utilization of cross-border transmission capacities while ensuring the physical constraints of the corresponding grids. Market coupling in Europe has been proposed since 2006 for creating a single European internal power market [29]. There are different ways of coupling markets, such as loose volume coupling, tight volume coupling and price coupling, which mainly consider economic aspects of a market-based mechanism [34]. In order to provide more flexibility to the grid, flow-based market coupling [49] has been advocated for highly distributed grids, as it takes essentially a more detailed grid model for transmission capacity determination into account. With the flow-based coupling approach, the power market is aware of the available transmission capacity in a more reliable way, since the transfers are based on real-time power flows rather than on predicted ones. In Central Western Europe (CWE), the first flow-based market coupling has been implemented in 2015 for the day-ahead market [50]. The deployment for the intraday market is still at the preparatory stage.

Another approach is the concept of a *real-time market* [51] that refers to a tool for system operators to balance power and/or congestion at any time during real-time operations of the transmission/distribution grid. According to the definition and practical relevance showed by Stoft [51], a real-time market should be linked with the balancing power market as an alternative option for the congestion management or intraday power re-dispatch. Stoft proposed a 5-minute time resolution for the real-time market, which indicates a more frequent update of the real-time pricing than within the intraday trading. This means also more frequent interactions with the physical grid for an update of the grid's state (e.g. generation and transmission capacity, congestion status, etc.), in order to establish a grid-dependent real-time pricing mechanism for minimizing the precision degradation of the power dispatch and grid regulation. In the EU project "EcoGrid EU" [12], the partners implemented this kind of real-time market as a direct control strategy of DER (Distributed Energy Resources) units, which complements the market-based mechanisms, such as the balancing power market. In the real-time market demo of "EcoGrid EU", authors [13] stated that the 5-minute time resolution holds a trade-off between the system complexity and the computational performance, which can be further shortened (e.g. 1-minute) for a more flexible and responsive power balancing, if the corresponding power generation and consumption can be measured at the same sampling rate.

In this dissertation, we employ a combination of market coupling and the real-time market (hereafter referred as "coupled real-time markets") for reliable intraday adjustments of the power dispatch of distributed power grids that represent for instance cross-border power systems among different regions. For modeling coupled real-time markets, we focus not on the bidding-and-auction process, but on the real-time pricing model. Following the proposed real-time market concept of EcoGrid EU [13], we propose coupled real-time markets as a balancing option that can be linked with the intraday or balancing market. In comparison with the Nordic power market, which has a regulating power market component that operates within the hour by the time of power delivery [43], the current German power market can only regulate a power imbalance either through the intraday trading (including 15-minute contracts) 30 minutes before the time of power delivery, or through the balancing reserves in different control stages after the time of power delivery. Therefore, our proposed coupled real-time markets serve as a regulating power market option to fill the gap in the balancing time scale, see Figure 5.

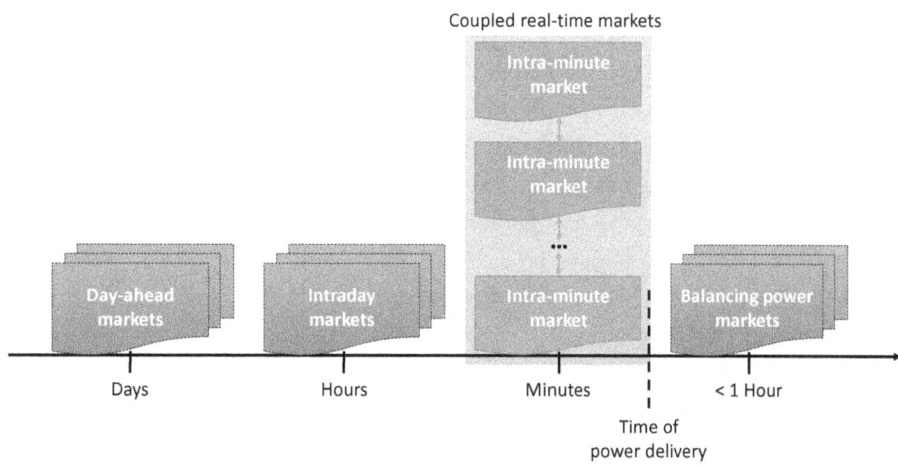

Figure 5.: The integration of coupled real-time markets as an intra-minute balancing option

Figure 5 shows the integration of proposed coupled real-time markets in the time scale among the existing power markets. As explained before, a price bidding takes place only in the day-ahead, intraday and balancing markets. After price determination on both the day-ahead and intraday markets, the real-time market proceeds with a real-time price settlement based on day-ahead market clearing prices or intraday adjusted market prices of the awarded orders (bids). Each individual real-time market can be considered as an extension of the intraday market in terms of the balancing capability with a finer time resolution. As seen in the figure, the time scale for the real-time market is defined with only a matter of minutes by the time of a physical delivery, which refers to one kind of "intra-minute market". Based on this intra-minute market, we develop a grid-dependent real-time price settlement model, in order to establish a market-grid coupling. From the market point of view, the market-grid coupling focuses on a real-time pricing signal that takes generation and consumption measurements as well as generation and transmission constraints into account, thereby reflecting the physical reality of the power grid. Since the objective of this dissertation is to develop a distributed market-grid coupling for distributed power grids, we model

45

for each power grid a correspondent intra-minute market (a real-time market) and apply market coupling among all intra-minute markets for a collaborative balancing service. The model description of this intra-minute power market with its real-time price settlement will be detailed in the next section.

3.2.1 Power Market Modeling

Since the launch of the 15-minute intraday trading on the German power market in December 2014, the intraday trading volumes[3] in the year of 2015 significantly increased by 42% from 26.4 TWh to 37.5 TWh, which shows highly dynamic behaviors on the German intraday market. Along with the trend of increasing system dynamics of a short-term power market, this dissertation provides a framework for developing dynamic models of such power markets than a market design solution. The aim of this section is to present the formulation of the proposed real-time market as a dynamic system that considers explicitly operational constraints of generation and transmission.

The research work on modeling power market as a dynamic system was pioneered by Alvarado [2]. By means of established differential equations that describe the market behavior under grid network constraints, Alvarado studied market equilibrium conditions in terms of price stability and volatility in consideration of network congestion. In the power market modeling survey of Ventosa et al. [54], this kind of modeling approach for representing market behavior considering competition is assigned to equilibrium models. The key task in the equilibrium-based power market modeling is to continuously maintain the equilibrium between generation and consumption, so that the demand can be balanced with the supply offer in each considered period of time while maximizing the benefits of market participants. The market equilibrium problems that are modeled in those equilibrium models, distinguish from each other with respect to competition models and strategic variables. In the existing literature [30, 27, 26, 11, 6, 35, 33, 47, 56], different oligopolistic competition models have been applied in power markets, among which the most widely used ones are the Cournot competition model (quantities as strategic variables), the Bertrand competition model (prices as strategic variables) and the Supply Function Equilibrium (SFE) model (both quantities and prices as strategic variables). These equilibrium models originally focus mainly on the supply side in terms of power generation scheduling of individual generation units. Although they take demand information into account, which is required for holding the supply-demand equilibrium, the optimization objective of these equilibrium models is formulated just for a profit maximization of suppliers as a single-sided pricing mechanism.

---

3 Trading statistics extracted from annual press release of EPEX SPOT in 2015

However, the most common auction procedure in the current power markets including EEX is the double-sided auction (Double Auction, DA), with which market participants can not only submit supply bits but also demand bits. In this case, for determining market prices as close to reality as possible, the dynamics of both supply and demand need to be taken into consideration. Along with the progressive development of Demand Side Management (DSM) programs [10, 42], real-time pricing as a market-oriented approach for DSM has been developed with increasing integration of demand dynamics [46]. This results in a paradigm shift of market equilibrium models that refer to a DA-based competition with suppliers and consumers.

Nicolaisen et al. [39] investigated the market power and efficiency implications of discriminatory DA pricing for power markets. Their investigation findings indicated that market efficiency is inherently considered in the design of the DA pricing mechanism and consumers as active market participants can limit the ability of suppliers to exercise market power. Kian et al. [31] applied the Nash-Cournot competition strategy in DA, in order to model power markets as dynamic systems with generating firms and load serving entities. In comparison with supplier-only auctions, their simulation results showed that double-sided auctions provide more market efficiency and price stability, and can limit the market power of suppliers. For the Chinese power market, Fang et al. [18] presented a DA model between large consumers and generators. Compared to the one-to-one transaction mode that was the state of the art in the Chinese power market, the proposed many-to-many transaction mode could improve market efficiency and optimize resource allocation in consideration of transaction and transmission costs. Also in the context of Demand Response (DR), Zugno et al. [59] presented a DA model based on the Stackelberg game between retailers and consumers. Different pricing schemes were studied for optimizing retailers' profits and consumers' flexible consumption concurrently. Recently, Taniguchi et al. [52, 53] proposed a DA-based DSM mechanism for a day-ahead power market with prosumers[4] as market participants. Authors developed a real-time pricing (RTP) mechanism based on submitted individual demand and supply functions of the prosumers, for maximizing the social welfare of the correspondent power grid.

Similar to the above DA-based market modeling approaches that focus mainly on the double-sided trading mechanism, our proposed intra-minute market employes as well a DA environment [14], in order to link it with the intraday market. As mentioned before, focus of the market modeling in this dissertation is rather a real-time price settlement model than a bidding-and-auction process. Within the following DA environment, we build optimization models for the real-time price

---

4 Prosumer: a person or an entity who generates and consumes power at the same time.

settlement among three market participant types including *Consumer*, *Producer* and *Network*.

Figure 6 shows a DA environment for the proposed intra-minute market in a sche-matic manner. The trading environment is based on the NOBEL[5] power market [28]. The NOBEL market model is composed of a series of overlapping "timeslots". Each timeslot corresponds to a time interval (e.g. 15 minutes) in the future that dictates when the traded electricity should be produced or consumed. Thus, participants trade based on their forecast levels of consumption and/or production. The time of the first timeslot (i.e. how close the market is to real-time trading), and the number of timeslots in the sequence (i.e. the maximum time horizon for the participants' forecasts) is configurable. Furthermore, the sequence is continuously updated on a rolling horizon by closing the nearest timeslot and opening a new one at the end of the sequence. Once a timeslot is closed, no further trading is allowed. Thus, the NOBEL market provides a common platform to enable power trading among smart grid stakeholders, such as residential, commercial and industrial consumers, conventional and renewable producers, as well as prosumers. For our intra-minute market, we extend the trading time interval — "time-slots" — dynamically subject to the sampling rate of the measurable or predictable power generation and consumption, which ranges from approximately 1 minute to 15 minutes.

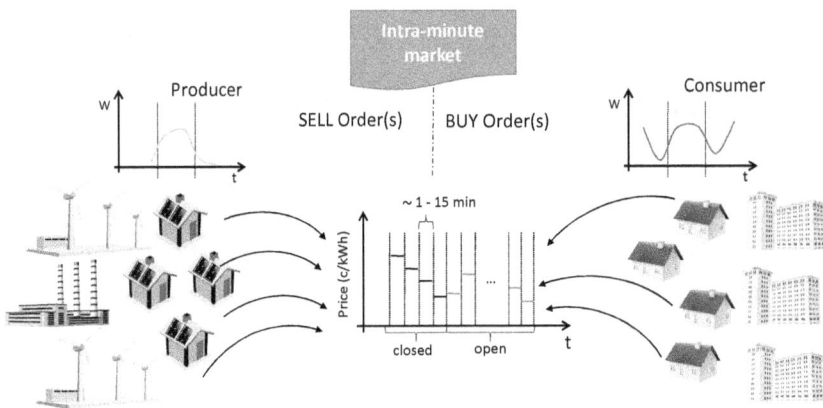

Figure 6.: Continuous-Double-Auction as trading mechanism for the proposed intra-minute market

---

5 NOBEL: Neighbourhood Oriented Brokerage Electricity and monitoring system, an EU FP7 project, www.ict-nobel.eu

The underlying trading mechanism in each timeslot is the continuous-double-auction (CDA). In a CDA, the market clears continuously; i.e. each time a new order is submitted, the market tries to match with the outstanding orders stored in a publicly viewable order book. This is in contrast to call auctions (CAs) that collect orders for a predetermined amount of time and clear at discrete time intervals. While in CAs the allocation is optimally computed by an auctioneer, in CDAs the allocation emerges from the continuous interaction between the participants. Although CDAs can lead to suboptimal outcomes, continuous allocation does provide an avenue for participants to adapt to dynamically changing market conditions. Due to the intermittent nature of mainstream renewable generation technologies, such as wind and solar, the dynamic behavior of household's demand, and how forecasts might change given exogenous information, the participants with CDAs have ample opportunity to update their standing on a timeslot as more information becomes available, or as market conditions change.

Generally, an order is composed of four values: timeslot, type (buy or sell), price and quantity. A transaction will occur whenever a buy and sell order agree in price, that is, the buy order price is greater or equal to a sell order price. If an order is unmatched, or only partially matched (it still has quantity left), the order is stored in the order book of the correspondent timeslot. The model also includes other order constraints that are accounted for by the matching process. For instance, an order can stipulate that its entire quantity must be met (i.e. no partial matching), or that any price will be accepted. For each timeslot, each participant will forecast consumption or production, determine its marginal cost (benefit) for selling (buying) electricity, and employ different strategies to maximize its economical outcomes. The pricing strategy for maximizing the social welfare among all market participants will be detailed in the next section. Moreover, this DA model has been shown to be both efficient [28] and scalable [23]. As an example, Fig. 7 depicts the trading outcomes for one day of market operation with 1897 participants, mainly households, of which 80% have solar production. The participants have a limit price for buying of $14\,c/kWh$, defined by their retailer contract (i.e. they will not pay more than what they already pay the retailer), and an assumed limit price of $5\,c/kWh$ for selling. This figure is based on results from the work of Goncalves Da Silva et al. [23].

In what follows, we present our real-time price settlement model for the intra-minute market model in this CDA environment. An important assumption is that all market participants are price takers, meaning no single one can influence prices unilaterally. Regarding the aforementioned physical and operational constraints of the power grid, not only *Consumer* and *Producer*, but also *Network* are considered in the following pricing model.

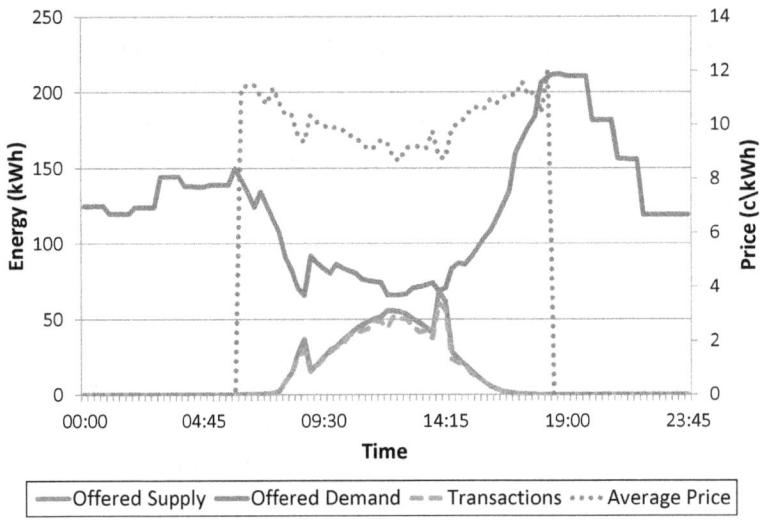

Figure 7.: The average traded price, transaction volume, and offered supply/demand of one day of trading between 1897 households of which 80% have solar production

3.2.2 *Market Formulation as a Dynamic System*

The intra-minute market model represents a local market for each grid unit that describes a single grid network in terms of a bus system. Details on power grid modeling in this dissertation will be provided in Section 3.3.2. The price settlement of the proposed local market is formulated as a dynamic market clearing among three market participant types: 1) *Consumer*, 2) *Producer* and 3) *Network*. The clearing mechanism that we follow is to determine the optimal clearing price through maximizing the total social welfare (or profits) of all market participants. In the following, we introduce the notation that allows the formulation of our proposed price settlement model for the intra-minute market.

Let $\mathbb{D}$, $\mathbb{S}$ and $\mathbb{T}$ denote generalized sets of consumers, producers and network players for each grid unit, respectively. For simplicity, we define an exemplary grid unit consisting of N bus nodes, indexed by $1, 2, \ldots, N$. Depending on the power flow state, each of these bus nodes can be classified as:

1. *Consumer node*, denoted as $d \in \mathbb{D}$ and $|\mathbb{D}| = N_D$, where there is only incoming power flow (i.e. demand);

2. *Generator node*, denoted as $s \in S$ and $|S| = N_S$, where there is only outgoing power flow (i.e. supply);

3. *Prosumer node*, denoted as $p \in \mathbb{P} = \mathbb{D} \cap S$ and $|\mathbb{P}| = N_P$, where there are both incoming and outgoing power flow (i.e. demand and supply);

where $N = N_D + N_S + N_P$. Since each prosumer node can contain both consumers and producers, the total number of consumers and producers is equal to or greater than the number of bus nodes within each grid unit. For simplicity, we define at each bus node a single "Consumer" or/and "Producer" for representing aggregate power consumption or/and generation at that node. This means that each consumer node has only one "Consumer", each generator node has only one "Producer" and each prosumer node has only one "Consumer" and one "Producer". Besides, we assume that in this grid unit power flow is linked from node to node by M transmission lines.

In order to enable the market-grid coupling, we follow the concept of locational marginal pricing (LMP) that refers to the theoretical power price at each bus node representing the marginal generation cost of that node for supplying an additional power unit while satisfying all the required constraints. So, for each trading time $t \geq 0$, a nodal price $\lambda_n(t)$ will be calculated as the market clearing price at each bus node $n \in \{1, 2, \ldots, N\}$. The price-taking assumption that we made before ensures that the nodal price $\lambda_n(t)$ cannot be influenced by the trading actions of consumers or suppliers, but be determined only by an optimization of their welfare functions. Before we describe the formulation of the price settlement model, we need to introduce the welfare function of each market participant type, which consists of both *utility* and *cost*.

Consumer: For time t at consumer or prosumer node $j \in \mathbb{D}$ or $\mathbb{P}$, we denote power demand by $D_j(t)$ and consumed power by $P_{D_j}(t)$, where $D_j(t) \geq P_{D_j}(t)$ should be satisfied for all t. If the power demand at node j at time t can be fully covered by sufficient generation of the grid unit, then there will be $D_j(t) = P_{D_j}(t)$. This implies that the consumer at node j at time t experiences a blackout or a load shedding, if there is $D_j(t) > P_{D_j}(t)$. We attach a utility function $u_j(x)$ and a disutility function $c_j(\Delta x)$ to the consumer at node j, which refer to an obtained monetary value from consuming $x = P_{D_j}(t)$ electricity power units and a monetary loss from shed load $\Delta x = D_j(t) - P_{D_j}(t)$, respectively. Besides the disutility cost, for the consumer, the main cost refers to the payment for the power consumption: $\lambda_j(t) \cdot P_{D_j}(t)$.

Producer: For time t at generator or prosumer node $i \in S$ or $\mathbb{P}$, we denote power generation for consumers by $P_{S_i}(t)$ and reserve generation for the balancing market by $R_{S_i}(t)$, where both $P_{S_i}(t)$ and $R_{S_i}(t)$ are subject to capacity limitations and ramping constraints of generators. The utility value of the producer is defined as the received revenue for producing $P_{S_i}(t)$ power units: $\lambda_i(t) \cdot P_{S_i}(t)$. In order to determine the welfare of the producer, we attach two cost functions $c_i^P(x)$ and $c_i^R(y)$ to the producer at node i, which refer to the generation cost at time t for injected power $x = P_{S_i}(t)$ and for reserve power $y = R_{S_i}(t)$, respectively.

Network: In order to capture the impact of physical constraints of the grid network on the market, network players are also modeled as market participants. Quite different from consumers and producers, network players neither generates nor consumes power. They are only responsible for the network constraints that consist of the supply-demand balance constraint $\sum_{j \in \mathbb{D} \bigcup \mathbb{P}} P_{D_j}(t) = \sum_{i \in \mathbb{S} \bigcup \mathbb{P}} P_{S_i}(t)$ and the transmission capacity limitations of the M transmission lines. The network welfare represents "toll charges" for the injected power at bus nodes as revenue of network players. For all network players, we assume that the power transmission can be conducted by him at zero cost, since the network infrastructure already exists. And the utility value of the network player $l \in \mathbb{T}$ at time t is defined as the received revenue for injecting power into bus node l: $\lambda_l(t) \cdot P_{in}^l(t)$, where $P_{in}^l(t)$ stands for the injected power quantity at bus node l at time t, which can be expressed as $P_{in}^l(t) = P_{D_l}(t) - P_{S_l}(t)$.

The welfare of market participants is defined as their profits from the market clearing. Based on the above description of market players, the welfare for an individual market participant can be then represented by the difference between the correspondent utility and cost. We denote the welfare of a consumer, a producer and a network player by $W_j(\lambda_j(t), x_j(t), \Delta x_j(t))$, $W_i(\lambda_i(t), x_i(t), y_i(t))$ and $W_l(\lambda_l(t), x_l(t))$, respectively.

$$W_j(\lambda_j(t), x_j(t), \Delta x_j(t)) \overset{\Delta}{=} u_j(x_j(t)) - c_j(\Delta x_j(t)) - \lambda_j(t) \cdot x_j(t), \quad (1)$$
$$where\ j \in \mathbb{D} \bigcup \mathbb{P},\ x_j(t) = P_{D_j}(t),\ \Delta x_j(t) = D_j(t) - P_{D_j}(t)$$

$$W_i(\lambda_i(t), x_i(t), y_i(t)) \overset{\Delta}{=} \lambda_i(t) \cdot x_i(t) - c_i^P(x_i(t)) - c_i^R(y_i(t)), \quad (2)$$
$$where\ i \in \mathbb{S} \bigcup \mathbb{P},\ x_i(t) = P_{S_i}(t),\ y_i(t) = R_{S_i}(t)$$

$$W_l(\lambda_l(t), x_l(t)) \overset{\Delta}{=} \lambda_l(t) \cdot x_l(t), \quad (3)$$
$$where\ l \in \mathbb{T},\ x_l(t) = P_{in}^l(t) = P_{D_l}(t) - P_{S_l}(t)$$

The operational goal in the price settlement model of our intra-minute power market is firstly to determine the nodal prices $\lambda_j(t)$, $\lambda_i(t)$ and $\lambda_l(t)$ for all correspondent bus nodes in the grid unit to maximize the total welfare of all market participants under all operational constraints. The total welfare that needs to be optimized is expressed as follows:

$$
W_{tot}(t) \stackrel{\Delta}{=} \sum_{j \in \mathbb{D} \cup \mathbb{P}} W_j(\lambda_j(t), x_j(t), \Delta x_j(t)) + \sum_{i \in \mathbb{S} \cup \mathbb{P}} W_i(\lambda_i(t), x_i(t), y_i(t)) \qquad (4)
$$
$$
+ \sum_{l \in \mathbb{T}} W_l(\lambda_l(t), x_l(t))
$$

We assume that all cost functions $c_j(\bullet)$, $c_i^P(\bullet)$ and $c_i^R(\bullet)$ are strictly convex and continuously differentiable, and the utility function $u_j(\bullet)$ is strictly concave and continuously differentiable. The above total welfare function $W_{tot}(t)$ shows mainly two input types, namely *nodal prices* and *power quantities*. However, from the market perspective, nodal prices are the only variables to be optimized. For convenience, let s_i and d_j be the inverse function of W_i and W_j for mapping the price to power generation and consumption, respectively.

$$
d_j(\lambda_j(t)) = W_j^{-1}(\lambda_j(t)) = \underset{x_j(t), \Delta x_j(t)}{\operatorname{argmax}} \ u_j(x_j(t)) - c_j(\Delta x_j(t)) - \lambda_j(t) \cdot x_j(t) \qquad (5)
$$
$$
s_i(\lambda_i(t)) = W_i^{-1}(\lambda_i(t)) = \underset{x_i(t), y_i(t)}{\operatorname{argmax}} \ \lambda_i(t) \cdot x_i(t) - c_i^P(x_i(t)) - c_i^R(y_i(t)) \qquad (6)
$$

Thus, consumers and producers can adjust their power consumption level $x_j(t)$, $\Delta x_j(t)$ and power production level $x_i(t)$, $y_i(t)$ based on the real-time nodal price information, respectively, i.e. according to the price-to-power mapping function $d_j(\lambda_j(t))$ (5) and $s_i(\lambda_i(t))$ (6). Therefore, set Equations (1), (2), (3), (5) and (6) into Equation (4), we can reformulate the total welfare function as the optimization model for a real-time nodal price settlement $\lambda_n(t)$ at each node n:

$$
\begin{aligned}
\max_{\lambda} W_{tot}(d(\lambda(t)), s(\lambda(t))) \stackrel{\Delta}{=} &\sum_{j \in \mathbb{D} \cup \mathbb{P}} \left(u_j(d_j(\lambda_j(t))) - c_j(d_j(\lambda_j(t))) - \lambda_j(t) \cdot d_j(\lambda_j(t)) \right) \\
&+ \sum_{i \in \mathbb{S} \cup \mathbb{P}} \left(\lambda_i(t) \cdot s_i(\lambda_i(t)) - c_i^P(s_i(\lambda_i(t))) - c_i^R(s_i(\lambda_i(t))) \right) \\
&+ \sum_{l \in \mathbb{T}} \lambda_l(t) \cdot (d_l(\lambda_l(t)) - s_l(\lambda_l(t))) \\
\stackrel{\Delta}{=} &\sum_{j \in \mathbb{D} \cup \mathbb{P}} \left(u_j(d_j(\lambda_j(t))) - c_j(d_j(\lambda_j(t))) \right) \qquad (7) \\
&- \sum_{i \in \mathbb{S} \cup \mathbb{P}} \left(c_i^P(s_i(\lambda_i(t))) + c_i^R(s_i(\lambda_i(t))) \right)
\end{aligned}
$$

subject to the operational constraints of the grid unit, which will be described in Section 3.3.2. The input variables $d = [d_1, \ldots, d_j, \ldots, d_n]^{tr}$ and $s = [s_1, \ldots, s_i, \ldots, s_n]^{tr}$ are the demand and supply vector representing power generation and consumption at individual bus nodes. $s_i = 0$ or $d_j = 0$ if the bus nodes i and j are not a prosumer node. The decision variable is then a vector of nodal prices $\lambda = [\lambda_1(t), \ldots, \lambda_n(t)]^{tr}$.

Since the welfare of network players is represented by the revenue for injecting power that takes place exactly at nodes, where consumers and produces are, we can see in Equation (7) that the network welfare is just compensated by consumers' payment and producers' revenue. We refer the determined nodal prices λ of the above optimization problem to the locational marginal costs for supplying an additional unit of load within a grid unit. By means of this nodal price based real-time pricing approach, we can already synthesize a market-grid coupling in a closed loop as seen in Figure 8.

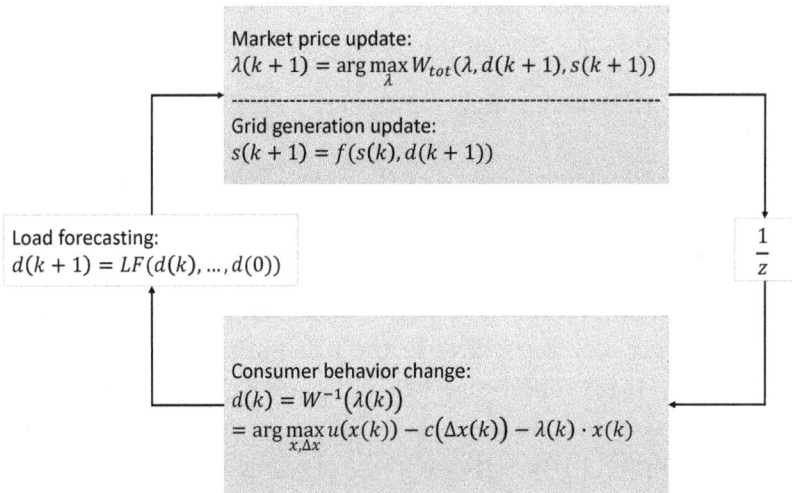

Figure 8.: A closed loop schema for market price update and grid generation update, where $\frac{1}{z}$ block represents the unit delay block specified by the sampling rate and k is therefore the correspondent discrete time step of t

Thus, based on a price-to-consumption mapping function $W^{-1}(\bullet)$ that describes the consumption behavior changes and a load forecasting function $LF(\bullet)$, both market price and power generation in the grid can be determined iteratively forward along the time axis. The market price is determined by the $W_{tot}(\bullet)$ optimization, while the grid generation update is conducted by a $f(\bullet)$ function of the generation optimization problem that will be formulated in the grid modeling section. However, Roozbehani et al. [45] demonstrated that this kind of market-grid feedback system could be unstable in terms of a high price volatility, if the market prices refer to LMP-based nodal prices. For this reason, a further operational goal in the price settlement model of the intra-minute power market is set to stabilize the determined nodal prices λ.

First, we introduce a local market price vector $\pi = [\pi_1(t), \ldots, \pi_n(t)]^{tr}$ that represents local intra-minute market prices at individual bus nodes. Then, we employ the dynamic pricing principle proposed by Roozbehani et al. [44] to calculate the local market prices π that stabilize the determined nodal prices λ. The pricing mechanism follows the calculation steps as below:

1. Discretize the time t with discrete time step k that corresponds to the time interval $[k, k+1]$, $k \in \mathbb{N}_0$;

2. For the dispatch time step k, assuming that the information of nodal prices, local market prices and nodal demands at the past T time steps are available, namely $\{\lambda(k-1), \ldots, \lambda(k-1-T)\}$, $\{\pi(k-1), \ldots, \pi(k-1-T)\}$ and $\{d(k-1), \ldots, d(k-1-T)\}$ are given;

3. Forecast each nodal demand d_j for the current time interval based on the past demand information: $\hat{d}_j(k) = LF_k(d_j(k-1), \ldots, d_j(k-1-T))$, $\forall j \in [1, n]$, where LF_k stands for a load forecasting function that can be time-variant;

4. Calculate nodal prices λ for the current time interval by solving (7) under predicted nodal demands $\hat{d}(k)$: $\lambda(k) = \text{argmax}_\lambda W_{tot}(\lambda | \hat{d}(k))$;

5. Calculate local market prices π for the current time interval with a pricing function Π_k: $\pi(k) = \Pi_k(\lambda(k), \ldots, \lambda(k-1-T), \pi(k-1), \ldots, \pi(k-1-T))$, where Π_k is a stabilizing function that describes dynamic relationships between nodal prices and local market prices;

6. Adjust each nodal load d_j for the current time interval based on the mapping function (5): $d_j(k) = d_j(\pi_j(k)) = W_j^{-1}(\pi_j(k)) = \text{argmax}_{x_j(k), \Delta x_j(k)} u_j(x_j(k)) - c_j(\Delta x_j(k)) - \lambda_j(t) \cdot x_j(k)$, $\forall j \in [1, n]$.

Thus, the complete price settlement model of the proposed intra-minute market is presented as shown in Figure 9. The price settlement model consists mainly of two operational goals, namely determining nodal prices and stabilizing local prices.

3.3 THE POWER GRID

A power grid is an interconnected network that is responsible for the power transport between power producers and consumers. Similar to other power grids in the world, a typical German power grid consists of two grid components, namely

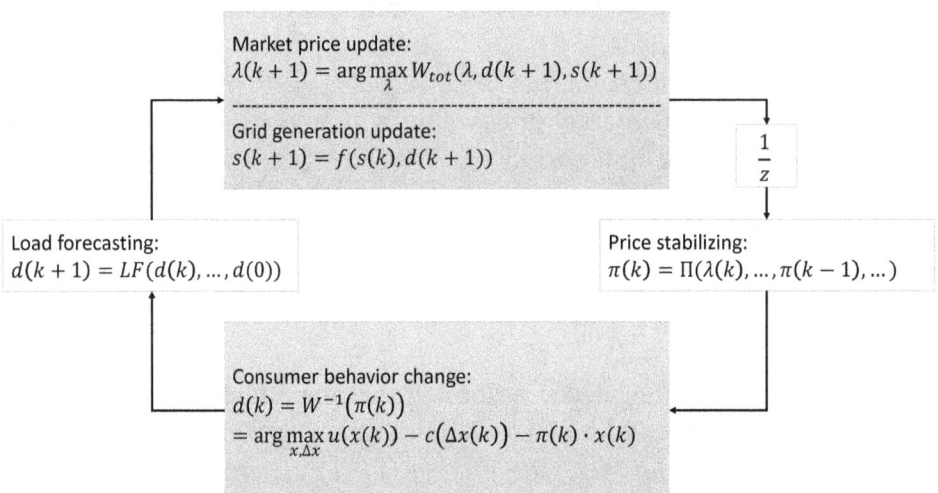

Figure 9.: A closed loop schema for market price update and grid generation update with a price stabilizing component

a transmission grid and a distribution grid, which covers four different voltage levels from extra high voltage (e.g. 230 kV) to low voltage (e.g. 230 V), see Figure 10. In the transmission grid as shown in Figure 10a, a bulk transmission of electricity from large power plants is conducted over long-distance high-voltage transmission lines to feed into the high voltage level for demand centers (wholesale consumers), while the distribution grid as shown in Figure 10b distributes electricity on the medium and low voltage level over distribution lines to individual end consumers. Large offshore wind farms are installed in the transmission grid, while distributed wind energy systems are commonly connected on the consumer side in the distribution grid.

As we know, the structure of a conventional power grid is primarily tailored to the bulk power generation and transmission in a centralized manner. The high penetration of renewables including distributed energy resources (DER) is increasing continuously, e.g. in Germany the renewable energy share of gross electricity consumption has an annual increase of at least 2-3% since 2010 and amounts to 27.4% in 2014 [8], which leads to a more variable residual load for the controllable power plants. A selective adaptation of the existing power grids can no longer

---

6 Adapted based on the figure from https://upload.wikimedia.org/wikipedia/commons/9/90/Electricity_Grid_Schematic_English.svg, "Electricity Grid Schematic English" by MBizon is licensed under CC BY 3.0.

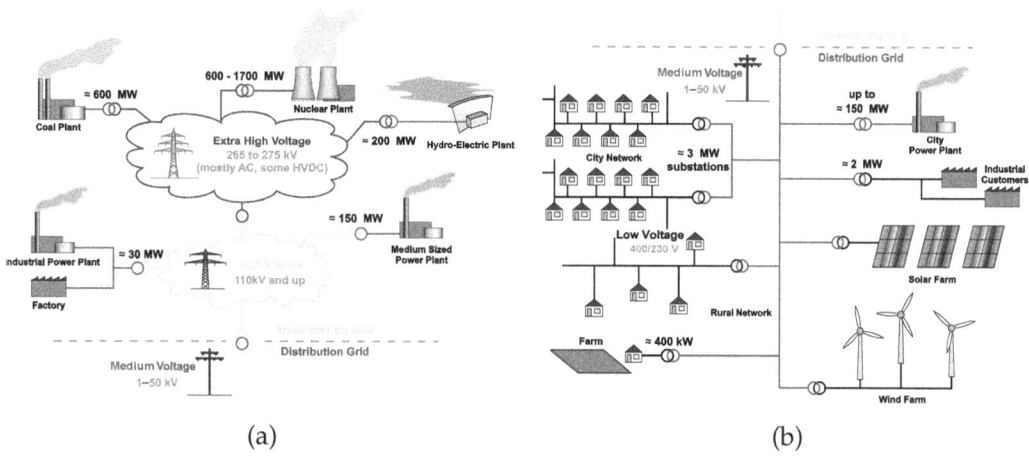

(a)   (b)

Figure 10.: Schematic representation of a power grid with different voltage levels[6]-
(a) Transmission grid network; (b) Distribution grid network

meet the requirement of an optimal demand-supply balance. In order to deal
with the grid integration of intermittent renewable energies and control decentralized power generation units, the existing power grids need to be updated towards
a new paradigm of smart grids [20]. The concept of smart grids focuses mainly
on the infrastructure and management system [19], such as two-way communication over advanced metering infrastructure (AMI) and demand-side management
(DSM) programs, which enables more progressive consumer-side applications. As
Figure 10 depicts, the existing power grid holds a strictly hierarchical architecture
from power generation over power transmission and distribution to power consumption. The smart grid exhibits rather a networked topology among all grid's
assets, see Figure 11. With the networked topology based on standard and interoperable communication protocols, grid components such as producer, consumer,
storage, transmission and distribution network are all interconnected and available for a necessary power re-dispatch in real time. Combined with IoT (Internet
of Things), smart grids are further developed towards the IoE (Internet of Energy)
concept [55], in which the information flow follows the power flow for collecting real-time information of each grid component with internet technologies, thus
establishing a virtual information grid in parallel with a physical power grid.

In addition, our proposed smart grid network, as seen in Figure 11, not only
bidirectionally interconnects grid components without the conventional hierarchy,
but also integrates the power market into the grid for enabling a market-grid
coupling. From the grid point of view, each market-grid coupling focuses on a
real-time re-dispatch signal that considers real-time price information from the
market for achieving an optimal economic dispatch. As stated in the Introduction

chapter, the goal of this dissertation is to develop a distributed market-grid coupling model for distributed power grids. Each individual grid unit is represented by a local smart grid as depicted in the above figure. Besides the described model of an intra-minute market in Section 3.2.2, we need additionally a grid model for the individual grid units, in order to model the proposed market-grid coupling. The following two sections introduce power grid modeling approaches in general and provides a detailed formulation of the power grid modeling employed in the dissertation. The model description of distributed power grids illustrates the relationships of the formalized grid components and variables. Based on the grid model description, we, then, formulate optimization frameworks for the grid operation towards an optimal economic dispatch. For the modeling of distributed grid units in dependence on the available grid data, we assume that either individual grid units (e.g. microgrids) have been extracted from a large grid network or a large grid network can be divided into several sub-grids (considered as grid units) with a certain disaggregation algorithm. Furthermore, the aforementioned physical and operational constraints in the market modeling in Section 3.2.2 will be detailed in the following grid model.

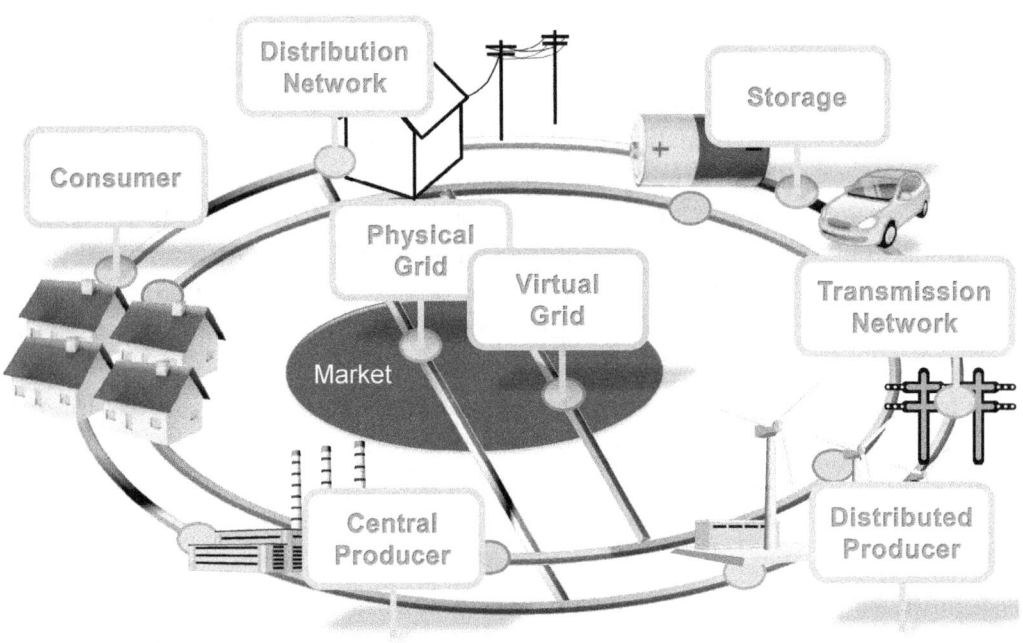

Figure 11.: A schematic representation of the smart grid topology

3.3.1 *Power Grid Modeling*

Power grid modeling is the base of grid system calculation, analysis and control. In order to study the power grid dynamics of a physical power system, the first step is to define the grid model, which requires hypotheses and simplifications [36]. As shown in both Figure 10 and 11, the components in a grid system are clearly representable; however, numerous research work in the field of power system analysis [48, 5, 38, 1, 40] have shown that relationships between the grid components and grid dynamics in general are more complex to model. For the purpose of a market-grid coupling in this dissertation, we need a grid model that describes not only the static part (grid topology and components) but also the dynamic part (grid constraints and power flows). Thus, the grid model employed in this dissertation consists of two abstraction layers: *Grid Topology Layer* and *Power Flow Layer*, see Figure 12.

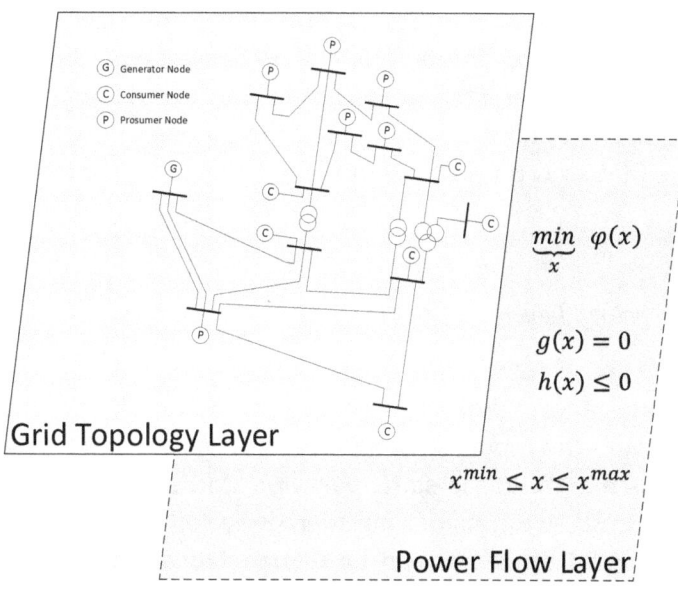

Figure 12.: Two-layer architecture of the proposed grid model

The one-line diagram is a common approach in the field of power engineering to represent the topology of a grid system (e.g. a bus system) and simplify grid components. Besides, a survey of Pagani and Aiello [41] showed that the notation of graph theory can be used to model topological properties of a power grid with a realistic representation of basic grid characteristics. In the grid topology layer, we follow the one-line diagram approach with a graph formulation for modeling the grid network as a directed graph. The main goal of this grid topology layer is to abstract the grid network infrastructure without affecting topological and physical properties of the grid and grid components. The modeling details of this layer is provided in Section 3.3.2.

The core of almost all power grid system representations in terms of grid dynamics is a set of equilibrium equations known as the power flow model [17]. This set of nonlinear differential algebraic equations (DAEs) is used to describe the grid system status and the entire power flow dynamics. Based on the equilibrium model of power flows, different optimization problems can be formulated for a grid control. In the power flow layer, we need a power flow model that analyzes either costs or prices of power flows, in order to couple market signals into the grid model. An optimal power flow (OPF) model [36, 17] is used to determine the minimum generation cost and loss, as well as the balance of the entire power flow at the same time. This optimization model takes into account not only the above mentioned power flow DAEs and physical grid constraints (e.g. transmission limits, active and reactive power limits, bus voltage limits, etc.), but also economic dispatch as an objective. Therefore, in the power flow layer, the power flow model employed in this dissertation focuses on an optimal power flow analysis. The formulation of the OPF-based power flow model for this layer is provided in Section 3.3.4.

3.3.2 The Grid Topology Layer

This section provides a formal description of a grid network as a graph. The grid graph model indicates the topology of a power grid network with its grid components, which refers mainly to the static part of a grid model. In order to model distributed power grids, we consider an interconnected power grid network consisting of $N_g \in \mathbb{N}$ grid units that can be interpreted as Microgrids or "Energy Hubs" [21]. Each grid unit has grid components, such as substations, generators, loads and network elements (e.g. transformers, transmission lines, phase shifters, etc.). For simplicity, we define each grid unit as a bus system, so that all components like substations, generators and loads are represented by bus nodes, while network elements are modeled with a branch model in terms of a standard π transmission line model [58, 4]. In a power flow study, the bus nodes are commonly classified into generator bus nodes or load bus nodes. In order to follow this concept, the previously defined three node types in the section of Power Market Modeling, will be merged into these two bus types, i.e. generator bus (generator and prosumer) and load bus (consumer). Branches that are used to connect the bus nodes within each grid unit and to interconnect the grid units are specified as intra-transmission and inter-transmission, respectively.

For the proposed grid graph model with N_g grid units, the whole model are formalized with the following notation:

1. Each grid unit $i \in [1, N_g]$ is described by a directed graph[7] $G^i = (V^i, E^i, A^i)$, where $V^i = \{v_1^i, \ldots, v_{n_i}^i\}$, $n_i \in \mathbb{N}$ is the set of n_i bus nodes, $E^i \subseteq V^i \times V^i$ is the set of directed edges representing intra-transmission branches, and $A^i \in \mathbb{R}^{n_i \times n_i}$ is a weighted adjacency matrix representing the nodal admittance matrix.

2. Directed edges are denoted as $e_{kl}^i = v_k^i \to v_l^i$ that means power flow in the branch e_{kl}^i from node v_k^i to node v_l^i.

3. The weights in the adjacency matrix are set as $[A^i]_{kk} = a_{kk}^i$ and $[A^i]_{kl} = -a_{kl}^i$, where a_{kk}^i is the self-admittance at the node v_k^i and a_{kl}^i is the admittance of the branch e_{kl}^i, if $e_{kl}^i \in E^i$; Otherwise $[A^i]_{kl} = 0$, if $e_{kl}^i \notin E^i$.

4. E^{ij} denotes the set of inter-transmission branches between the grid unit G^i and G^j.

5. Ω_l^i defines the set of bus nodes connected to node v_l^i within G^i, which refer only to the bus nodes connected by intra-transmission branches. Ω_{G^i} defines the set of interconnections to and from G^i.

6. Θ^i denotes the vector of voltage angles $\{\theta_1^i, \ldots, \theta_{n_i}^i\}$, and θ_{kl}^i is the phase angle difference $\theta_k^i - \theta_l^i$ between node v_k^i and v_l^i.

7. V^i and I^i are denoted as a complex vector of node voltages and node current injections within the grid unit G^i, respectively.

8. $P_{n,g}^i$, $P_{n,l}^i$, $P_{n,in}^i$, $Q_{n,g}^i$, $Q_{n,l}^i$ and $Q_{n,in}^i$ denote the active generator power, active load power, active net power injection, reactive generator power, reactive load power and reactive net power injection at the bus node v_n^i, respectively.

9. P_{in}^i, Q_{in}^i, P_{out}^i and Q_{out}^i are defined as active and reactive power flows into and out of the grid unit G^i, respectively.

In this dissertation, we mainly work on the bus models of IEEE, which are available open model cases of synthetic grid samples. As an example, we demonstrate how the IEEE 300 bus test case[8] can be modeled as the proposed grid graph. Bus

---

7 In general, we use upper indices for the notation of grid units and bottom indices for the notation within each individual grid unit.

8 http://www.ee.washington.edu/research/pstca/

data, branch data and generator data in the test case are used to specify the grid graph. Figure 13 shows an overview of the grid graph of the IEEE 300 bus test case, which is generated by the graph database Neo4j. In this graph, both generator and load bus nodes are represented by orange nodes. All 69 generators are denoted by green nodes that are connected to generator bus nodes. Furthermore, all 400 branches are denoted by directed gray edges.

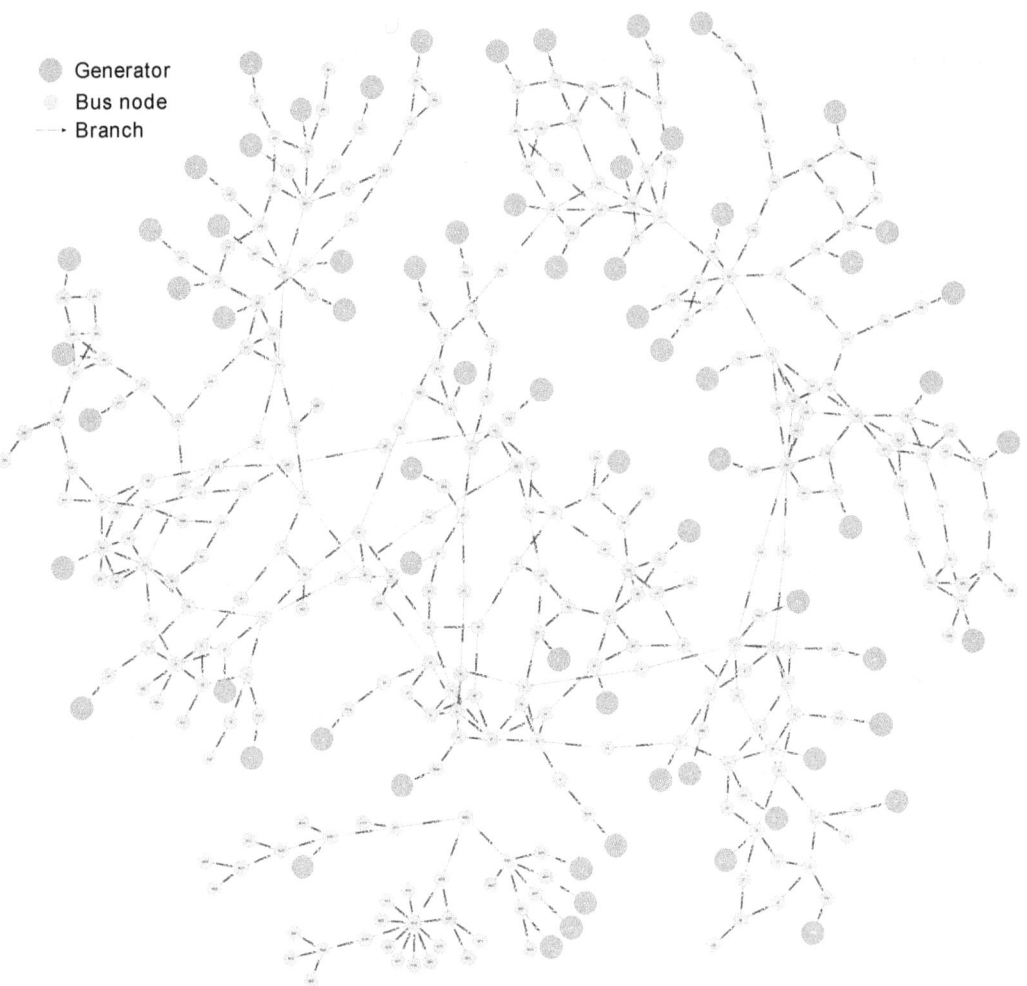

Figure 13.: An overview of the generated grid graph by the graph database Neo4j for the IEEE 300 bus test case

3.3.3 *Grid Disaggregation*

For the model of distributed grid units, we assumed in the previous section that either individual grid units have already been extracted from a large grid network or a large grid network can be divided into several interconnected grid units. In order to generate a model of distributed grid units out of the IEEE 300 bus test case, a grid disaggregation algorithm that is employed in this dissertation will be introduced based on the given grid graph regarding grid physical and operational properties (e.g. generation capacity, transmission limitation, etc.). First of all, the definitions of *Path* and *Path Length* as well as *Shortest Path* between two nodes are required to understand the disaggregation algorithm below.

Path and Path Length: A *path* of the directed graph G between the node v_k and v_l is a sequence of edges to connect both nodes as end-nodes, where all edges are directed in the same direction. The *path* has the form of a subgraph $P = (V(P), E(P))$, where $V(P) = \{v_k, v_{k+1}, \cdots, v_{l-1}, v_l\}$ and $E(P) = \{(v_k, v_{k+1}), (v_{k+1}, v_{k+2}), \cdots, (v_{l-1}, v_l)\} = \{e_{k,k+1}, e_{k+1,k+2}, \cdots, e_{l-1,l}\}$. Then, the correspondent *path length* l_P is defined by the number of edges in the edge set $E(P)$: $l_P = |E(P)|$.

Shortest Path: The *shortest path* between the node v_k and v_l is the *path* with the minimum *path length*. Let the set $\{P_1, P_2, \cdots, P_n\}$ be all possible paths between v_k and v_l, then the *shortest path* P_s is defined by $\mathrm{argmin}_P(\{|P_1|, |P_2|, \cdots, |P_n|\})$.

In order to reflect the reality of a power grid, particularly its transmission constraints in terms of the transmission capacity at branches, we adapt the *path length* to a *weighted path length*. Based on the defined adjacency matrix in the grid topology layer, we attach the admittance of branches as weights to edges. The above defined path length l_P is adapted by the sum of weights of correspondent edges: $l_P^w = \sum \{a_{k,k+1}, a_{k+1,k+2}, \cdots, a_{l-1,l}\}$. Then, shortest paths will be determined based on l_P^w. Furthermore, we denote the number of all shortest paths between the node v_k and v_l by σ_{kl}, and the number of shortest paths that pass through the edge e by $\sigma_{kl}(e)$.

Then, we propose a grid disaggregation that refers to a division of the whole bus nodes, in order to disaggregate the large grid network into several grid units regarding transmission limitations, such as power losses. From the branch point of view, the disaggregation mechanism needs to determine the importance of each branch. In graph theory, *betweenness centrality* of an edge can be used as a measure to describe the importance of the edge with respect to shortest paths. A formal definition the *edge betweenness* is provided as follows:

Edge Betweenness: Edge betweenness is a centrality measure that quantifies the number of times an edge is passed through by shortest paths between two nodes. Mathematically, the edge betweenness of an edge e between the node v_k and v_l is expressed as the sum of the fraction between σ_{kl} and $\sigma_{kl}(e)$:

$$c_B(e) = \sum_{v_k, v_l \in V} \frac{\sigma_{kl}(e)}{\sigma_{kl}} \qquad (8)$$

Listing 3.1: Grid Disaggregation Algorithm

```
targetClusterCount = <set>;
clusterCount = graph.getClusterCount();
validGraph.push_back(graph);
clustering: for(int i = 0; i < iterationNumber; ++i) {
  vectorE = graph.getEdges();
  // request all edges of a graph
  centralityVector = C_B(vectorE);
  centriestE = max(centralityVector);
  // find the edge with the highest betweenness centrality
  graph.remove(centriestE);
  if(clusterCount < graph.getClusterCount()) {
    // find graph with higher clusterCount
    clusters = graph.getClusters();
    foreach(cluster : clusters){
      if(cluster.getGeneratorCount() == 0)
      // at least one cluster doesn't contain a generator
        break clustering; // therefore break the loop
    }
    clusterCount = graph.getClusterCount();
    validGraph.push_back(graph);
    if(targetClusterCount == clusterCount)
      break clustering;
  }
}
```

Our grid disaggregation algorithm is based on the Girvan-Newman algorithm [22] that works as a clustering approach based on the definition and notation of the edge betweenness. The Girvan-Newman algorithm proposed by Girvan and Newman is used to detect communities by iteratively removing edges. In each iteration, the algorithm removes the edge, through which the highest number of shortest paths pass by. Thus, the algorithm tries to separate inter-connection nodes between edges. Based on the clustering result of the Girvan-Newman algorithm, the grid disaggregation algorithm performs with a given target number of grid units (`targetClusterCount`) as described in the pseudo code in Listing 3.1. First, the algorithm divides the given grid network into maximal possible grid units (`getClusters()`), so that each individual grid unit has a high network density in terms of intra-transmission branches. Then, it checks further if each grid unit (`cluster`) contains at least one generator node with the method `getGeneratorCount()`, in order to determine valid grid units as a final step.

By means of the above grid disaggregation algorithm, the grid graph of the IEEE 300 bus test case is separated into 4 grid units (G^1, G^2, G^3 and G^4), where bus nodes are electrically close to each other within each grid unit with respect to power losses. All bus nodes with branches are split up into correspondent grid units, see Table 1. The partitioning on the grid graph of the IEEE 300 bus test case is depicted in Figure 14.

Table 1.: The number of bus nodes, branches and interconnections of individual grid units

Grid Units	# Generator Bus	# Load Bus	# Branch	# Interconnection to G^1	G^2	G^3	G^4
G^1	26	96	166	-	3	6	-
G^2	22	58	114	-	-	-	-
G^3	16	47	83	1	-	-	-
G^4	5	30	37	-	-	-	-

3.3.4 *The Power Flow Layer*

In the power flow layer, we introduce for each individual grid unit an OPF-based power flow model that optimizes the steady-state performance of the correspondent grid unit in terms of minimizing generation cost, loss, etc., while satisfying power flow equality and inequality constraints as well as other generation and transmission limitations. This layer refers mainly to the dynamic part of a grid model. Regarding the notation, the OPF problem for each grid unit can be formulated in the same form as an optimization problem. For simplicity, we describe the OPF formulation for a general grid unit as follows:

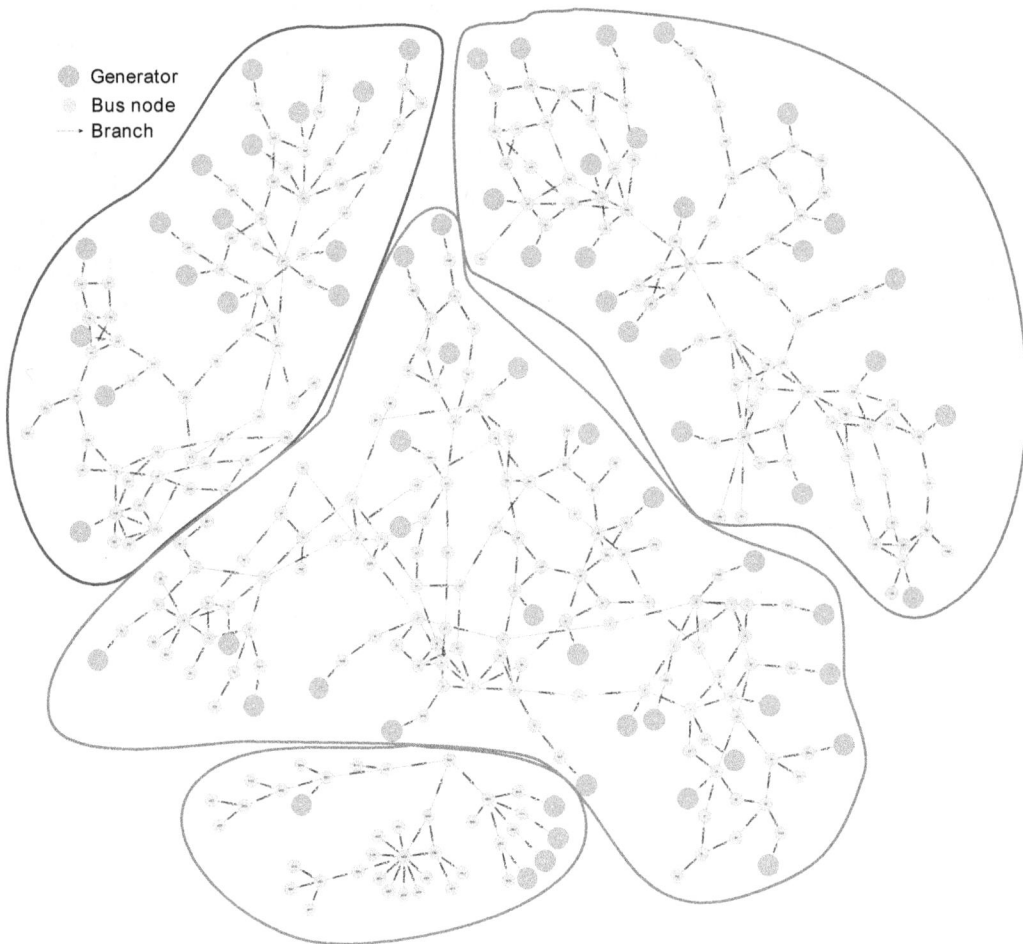

Figure 14.: An overview of the grid disaggregation into 4 grid units: the gray-highlighted grid unit refers to G^1; the green-highlighted grid unit refers to G^2; the blue-highlighted grid unit refers to G^3; the red-highlighted grid unit refers to G^4

$$\min_{x} \quad \varphi(x) \tag{9}$$
$$\text{subject to} \quad g(x) = 0$$
$$h(x) \leq 0$$
$$x^{min} \leq x \leq x^{max}$$

where the optimization variable $x \in \mathbb{R}^{n_x}$ is defined in terms of a $n_x \times 1$ vector of bus voltage angles Θ and magnitudes V as well as real and reactive powers P and Q of bus nodes. x^{min} and x^{max} represent the variable limits of each optimization variable. $\varphi(x)$ is the objective function ($\varphi(x) : \mathbb{R}^{n_x} \mapsto \mathbb{R}$), $g(x)$ are equality constraints ($g(x) : \mathbb{R}^{n_x} \mapsto \mathbb{R}^{n_g}$), and $h(x)$ are the inequality constraints ($h(x) : \mathbb{R}^{n_x} \mapsto \mathbb{R}^{n_h}$), both $n_g, n_h < n_x$.

Regarding the power flow we follow again the classification and notation of bus nodes as in Section 3.2.2: for a N-bus grid unit, there are N_S generator nodes denoted by $s \in \mathbb{S}$, N_D consumer nodes denoted by $d \in \mathbb{D}$ and N_P prosumer nodes denoted by $p \in \mathbb{P}$, where $N = N_G + N_C + N_P$. Besides, we assume M branches to transport power from node to node. The optimization variable x for the proposed power flow model with these three bus types is then extended by $x = \begin{bmatrix} \Theta, V, P_s, Q_s, P_p, Q_p, P_d \end{bmatrix}^{tr}$. In order to consider the controllability of the power grid, we introduce u as a vector of additional independent variables that represent the controllable quantities in the system, such as transformer tap settings, shunt VAR compensations, etc. Considering the power flow equations and transmission limitations among these N bus nodes, the OPF-based power flow model can be formulated as an optimization problem in (10): where the objective function $\varphi(x; u)$ is a summation of individual cost functions of generator powers ($f_s(P_s)$), consumer powers ($f_d(P_d)$) and prosumer powers ($f_p(P_p)$). The details of the individual cost functions are described in Section 3.2.2. The equality constraints consist of two sets of N non-linear nodal power balance equations (generator powers = load powers + injected powers), one for real powers g_P and the other one for reactive powers g_Q, in which P_s, Q_s, P_p, Q_p, P_d stand for the generator real and reactive powers, the prosumer real and reactive powers, and the consumer load powers, respectively. The inequality constraints ϕ_{ij} and ϕ_{ji} represent M flow limits of the active powers flowing through the branches in both directions. The variable limits include upper and lower bounds of generation outputs (P_s and Q_s; possibly P_p and Q_p), power loads (P_d; possibly P_p), stability or security limits (V and Θ) and controllability (u, such as transformer constraints, etc.).

$$\min_{x;u} \quad \varphi(x;u) = \sum_{s \in S} f_s(P_s) + \sum_{d \in \mathbb{D}} f_d(P_d) + \sum_{p \in \mathbb{P}} f_p(P_p) \qquad (10)$$

$$\text{subject to} \quad g_P(\Theta, V, P_s, Q_s, P_p, Q_p, P_d; u) = 0$$

$$g_Q(\Theta, V, P_s, Q_s, P_p, Q_p, P_d; u) = 0$$

$$\left| \phi_{ij}(\Theta, V) \right| \leq \phi_{ij}^{max}$$

$$\left| \phi_{ji}(\Theta, V) \right| \leq \phi_{ji}^{max}$$

$$P_s^{min} \leq P_s \leq P_s^{max}$$

$$Q_s^{min} \leq Q_s \leq Q_s^{max}$$

$$P_d^{min} \leq P_d \leq P_d^{max}$$

$$P_p^{min} \leq P_p \leq P_p^{max}$$

$$Q_p^{min} \leq Q_p \leq Q_p^{max}$$

$$V^{min} \leq V \leq V^{max}$$

$$\Theta^{min} \leq \Theta \leq \Theta^{max}$$

$$u^{min} \leq u \leq u^{max}$$

To solve the above optimization problem, we consider the Lagrange dual function $\mathcal{L}$ associated to the OPF problem in (10) by employing a vector of slack variables s. With the Lagrange dual function, we can then transform inequality constraints into equalities as expressed in (11): where ρ_P and $\rho_Q \in \mathbb{R}^N$, and all the $\lambda > 0$ are the Lagrange multipliers. The s variables are the individual non-negative slack variables used to transform the inequality constraints to equalities. Both μ_{min} and μ_{max} are barrier parameters for the logarithmic barrier function of the slack variables. In the next section, we first illustrate how to use this proposed power flow model to determine the nodal prices that are introduced in Section 3.2.2; then introduce a feedback modeling approach to combine both the market and grid model for a formal definition of the market-grid coupling.

$$\mathcal{L} = \sum_{s \in \mathbb{S}} f_s(P_s) + \sum_{d \in \mathbb{D}} f_d(P_d) + \sum_{p \in \mathbb{P}} f_p(P_p) \tag{11}$$
$$- \rho_P^\top g_P(\Theta, V, P_s, Q_s, P_p, Q_p, P_d; u)$$
$$- \rho_Q^\top g_Q(\Theta, V, P_s, Q_s, P_p, Q_p, P_d; u)$$
$$- \lambda_{\phi_{ij}^{max}}^\top (\phi_{ij}^{max} - \phi_{ij}(\Theta, V) - s_{\phi_{ij}^{max}})$$
$$- \lambda_{\phi_{ji}^{max}}^\top (\phi_{ji}^{max} - \phi_{ji}(\Theta, V) - s_{\phi_{ji}^{max}})$$
$$- \lambda_{P_s^{max}}^\top (P_s^{max} - P_s - s_{P_s^{max}})$$
$$- \lambda_{P_s^{min}}^\top (P_s - P_s^{min} - s_{P_s^{min}})$$
$$- \lambda_{Q_s^{max}}^\top (Q_s^{max} - Q_s - s_{Q_s^{max}})$$
$$- \lambda_{Q_s^{min}}^\top (Q_s - Q_s^{min} - s_{Q_s^{min}})$$
$$- \lambda_{P_d^{max}}^\top (P_d^{max} - P_d - s_{P_d^{max}})$$
$$- \lambda_{P_d^{min}}^\top (P_d - P_d^{min} - s_{P_d^{min}})$$
$$- \lambda_{P_p^{max}}^\top (P_p^{max} - P_p - s_{P_p^{max}})$$
$$- \lambda_{P_p^{min}}^\top (P_p - P_p^{min} - s_{P_p^{min}})$$
$$- \lambda_{Q_p^{max}}^\top (Q_p^{max} - Q_p - s_{Q_p^{max}})$$
$$- \lambda_{Q_p^{min}}^\top (Q_p - Q_p^{min} - s_{Q_p^{min}})$$
$$- \lambda_{V^{max}}^\top (V^{max} - V - s_{V^{max}})$$
$$- \lambda_{V^{min}}^\top (V - V^{min} - s_{V^{min}})$$
$$- \lambda_{\Theta^{max}}^\top (\Theta^{max} - \Theta - s_{\Theta^{max}})$$
$$- \lambda_{\Theta^{min}}^\top (\Theta - \Theta^{min} - s_{\Theta^{min}})$$
$$- \lambda_{u^{max}}^\top (u^{max} - u - s_{u^{max}})$$
$$- \lambda_{u^{min}}^\top (u - u^{min} - s_{u^{min}})$$
$$- \mu_{min} \sum_i \ln s_{i^{min}} - \mu_{max} \sum_i \ln s_{i^{max}}$$

3.4 THE MARKET-GRID COUPLING

In principle, a coupling for the market model and the grid model can be designed by means of two approaches, i.e. 1) a system integration approach that models the coupling as an extended system modeling with a market component and a grid component; 2) a feedback modeling approach that models the coupling as an information processing step between the existing market and grid model. The system integration approach requires an extended system model that considers the market model and the grid model as inherent system components. That means in this extended system model, the market and the grid will be formulated jointly by a set of dependent parameters. In practice, this approach is not feasible due to the complexity of the individual market and grid models. In this dissertation, we selected the feedback modeling approach as a coupling concept. In order to investigate the interoperable controllability between the market and the grid within a market-grid coupling framework, we focus intuitively on a feedback control concept for modeling the market-grid coupling.

In this section, we formalize a feedback control model to describe the market-grid coupling. The interpretation of a market-grid coupling in this dissertation is twofold. From a grid point of view, the optimum value of reliability in power supply is an instantaneous power balance based on the deregulated power market, which implies the balance between customers' marginal increase and power flow transmission cost [3]. From a market point of view, the power transmission system should be simplified towards a system to inject and withdraw the traded power [7]. Therefore, **the definition of our proposed market-grid coupling refers to a dynamic interaction between the physical power grid and the economic power market, which focuses rather on a bi-directional economic dispatch problem than a classical one**.

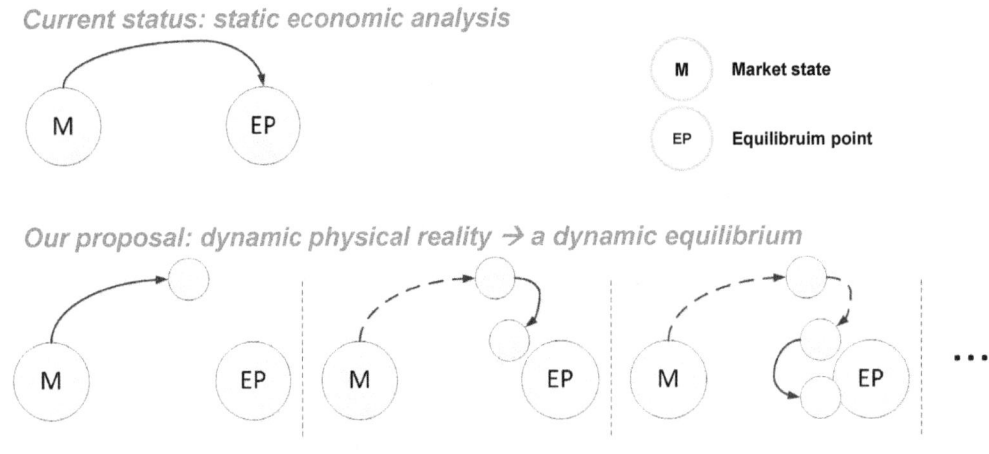

Figure 15.: A dynamic equilibrium for the proposed market-grid coupling

Figure 15 depicts schematically the development of a market equilibrium, where M circle stands for the market state and EP circle stands for an supply-demand equilibrium point. The supply-demand equilibrium point of the most current power market models is an end point in an economic analysis of standard economic models. The goal of our proposed market-grid coupling is to bring the dynamic physical reality of the power grid into the market equilibrium, in order to extend the usual competitive equilibrium (producers and consumers only) towards a dynamic competitive equilibrium among producers, consumers and network players. As the coupling approach, we employ feedback control theory to model the market-grid coupling.

In order to establish a feedback control model, we need first to define the feedback signal. The nodal price λ that is introduced in Section 3.2.2 can serve as such a proper feedback signal, since it refers to the theoretical price of electricity at each node in the power grid, which combines the grid reality and market price. In the following, we illustrate how to determine the nodal price and based on that couple the market and grid in a control loop.

The nodal price λ can be calculated through locational marginal pricing (LMP) within an OPF framework, which is then defined as the marginal cost to supply an additional unit of load at that node while satisfying all the required constraints. According to the LMP decomposition into marginal energy price, marginal congestion price and marginal loss price [32, 57], the nodal price λ at node $i \in 1, 2, \ldots, N$ can be calculated based on the determined Lagrange multipliers of the Lagrange dual function (11) as follows:

$$\lambda_i = \lambda_i^e + \lambda_i^l + \lambda_i^c \tag{12}$$

$$\lambda_i^e = \begin{bmatrix} \rho_{Pr} \\ \rho_{Qr} \end{bmatrix} \tag{13}$$

$$\lambda_i^l = - \left(1 - [\mathcal{J}_m]^{-1} \mathcal{J} \begin{bmatrix} 1 - \frac{\partial P_L}{\partial P_{i,in}} & -\frac{\partial Q_L}{\partial P_{i,in}} \\ -\frac{\partial P_L}{\partial Q_{i,in}} & 1 - \frac{\partial Q_L}{\partial Q_{i,in}} \end{bmatrix} \right) \begin{bmatrix} \rho_{Pr} \\ \rho_{Qr} \end{bmatrix} \tag{14}$$

$$\lambda_i^c = [\mathcal{J}_m]^{-1} \mathcal{J} \sum \begin{bmatrix} \frac{\partial h(x)}{\partial P_{i,in}} \\ \frac{\partial h(x)}{\partial Q_{i,in}} \end{bmatrix} \left(\lambda^{max} - \lambda^{min} \right) \tag{15}$$

where ρ_{Pr} and ρ_{Qr} correspond to the Lagrange multipliers of the real and reactive power balance equation at the reference node (slack bus). P_L and Q_L are system real and reactive power loss, while $P_{i,in}$ and $Q_{i,in}$ represent nodal injection real and reactive power at node i. $h(x)$ stands for a vector of all the inequality transmission constraints and λ^{max}, λ^{min} are correspondent Lagrange multipliers of the transmission constraints. Both Jacobian matrices $\mathcal{J}$ and $\mathcal{J}_m$ can be calculated as follows:

$$\mathcal{J} = - \begin{bmatrix} \frac{\partial P_{i,in}}{\partial V_i} & \frac{\partial Q_{i,in}}{\partial V_i} \\ \frac{\partial P_{i,in}}{\partial \Theta_i} & \frac{\partial Q_{i,in}}{\partial \Theta_i} \end{bmatrix} \tag{16}$$

$$\mathcal{J}_m = - \begin{bmatrix} \frac{\partial P_{i,in}}{\partial V_i} + \frac{\partial P_{i,l}}{\partial V_i} & \frac{\partial Q_{i,in}}{\partial V_i} + \frac{\partial Q_{i,l}}{\partial V_i} \\ \frac{\partial P_{i,in}}{\partial \Theta_i} + \frac{\partial P_{i,l}}{\partial \Theta_i} & \frac{\partial Q_{i,in}}{\partial \Theta_i} + \frac{\partial Q_{i,l}}{\partial \Theta_i} \end{bmatrix} \tag{17}$$

where V_i and Θ_i are the voltage magnitude and angle at node i. Since we consider dynamic power demand transactions (only active load powers), $P_{i,l} = P_{i,d} \vee P_{i,p}$ and $Q_{i,l} = 0$ represent the real and reactive power loads at consumer or prosumer node i.

Finally, we calculate $\lambda_i^P - \lambda_j^P$ (P indicates the real part that refers to the real power) as the real power transaction charges from node j to node i and generate a dynamic coefficient matrix C for a real-time update of the power market:

$$
C = \begin{bmatrix} c_{11} & c_{12} & \cdots & c_{1N} \\ c_{21} & c_{22} & \cdots & c_{2N} \\ \vdots & \vdots & \ddots & \vdots \\ c_{N1} & c_{N2} & \cdots & c_{NN} \end{bmatrix} \tag{18}
$$

$$
= \begin{bmatrix} \dfrac{|\lambda_1^P - \lambda_1^P|}{\lambda_1^P} & \cdots & \dfrac{|\lambda_1^P - \lambda_N^P|}{\lambda_N^P} \\ \vdots & \ddots & \vdots \\ \dfrac{|\lambda_N^P - \lambda_1^P|}{\lambda_1^P} & \cdots & \dfrac{|\lambda_N^P - \lambda_N^P|}{\lambda_N^P} \end{bmatrix} \tag{19}
$$

$$
= \begin{bmatrix} 0 & \cdots & \dfrac{|\lambda_1^P - \lambda_N^P|}{\lambda_N^P} \\ \vdots & \ddots & \vdots \\ \dfrac{|\lambda_N^P - \lambda_1^P|}{\lambda_1^P} & \cdots & 0 \end{bmatrix} \tag{20}
$$

Coupling Requirements

Based on the above coefficient matrix C, a feedback control loop can be formulated for coupling the market and grid. Regarding individual components in the feedback control loop, coupling requirements in terms of feedback signals are depicted in Figure 16. As shown in the figure, the market-grid coupling is simplified as a basic control loop, including a plant, a controller and an additional transducer. The plant is the final object under control and refers to the proposed OPF-based power grid model. Thus, the state space, which describes the plant, consists of real and reactive power state variables as well as nodal voltage magnitude and angle variables. The transducer is used to monitor the state variables of the plant. In this control loop, the transducer performs the dynamic locational marginal pricing (DLMP) algorithm (an iterative calculation of LMP-based nodal prices in Equation (12)) to produce observed relative transaction charges between each two nodes in terms of a coefficient matrix C. The reference transaction charges are compared to the calculated C matrix (feedback signal) to generate an error signal, which implies at this point the possibility of an update of the market clearing. This error signal is fed into the controller block, which employs the intra-minute market model to settle a local market price that augments the nodal price with the effective power transmission costs between nodes, and updates the actual production/consumption for each node in terms of time series, which are then fed back into the plant for the next run of OPF execution.

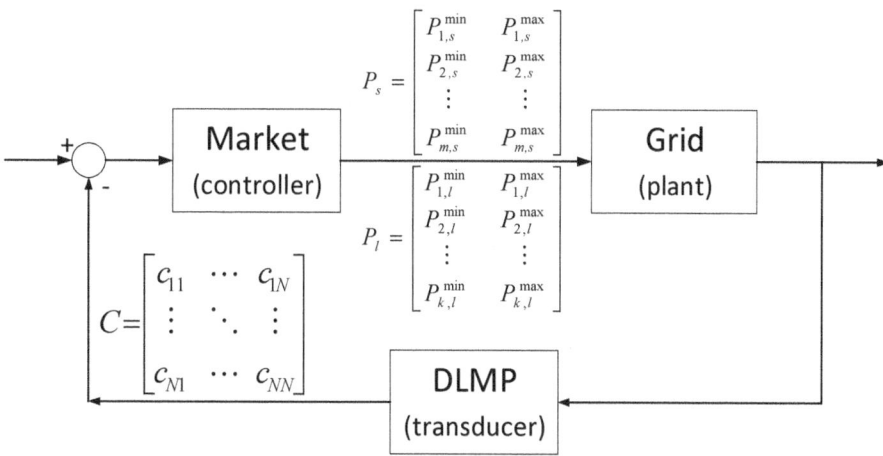

Figure 16.: A control loop for real-time interaction between the power market and grid

Moreover, the control hierarchy of a power grid consists of four different levels [24], including *grid planning and expansion, generation unit commitment, power flow optimization*, and *voltage or frequency control*. Regarding the purpose of the proposed market-grid coupling as well as its response time, the control task of this market-grid coupling is allocated at the control hierarchy level of a power flow optimization, which works on a time scale of minutes. The intra-minute market model as presented in Section 3.2 is designed for a price settlement on a time scale of minutes, which meets inherently the time response requirement. Then, the open question regarding the coupling requirement is whether the augmented power flow optimization model based on this market-grid coupling can be solved within the same time scale and provides high scalability for large grid networks. We will address this question in the next chapters.

Initial Test

As an initial test on this market-grid coupling model, we have performed a number of preliminary experiments with the intra-minute market to verify the clearing algorithm. We have simulated the market operating on a power grid system based on the IEEE 14-bus test case with 14 prosumer households. Market operation was simulated during one timeslot (15 minutes) in which half of the prosumers acted as buyers, and the other half as sellers. We have compared the transactions generated by the market with the power system constraints incorporated and without them. The coefficient matrix was set up such that it would be cost prohibitive to have electricity transfers between certain nodes. When the power transmission constraints were not incorporated, 100% of the available power on the market was traded. With constraints in the coefficient matrix, the traded energy dropped to 89%, as some generators were unable to trade due to power flow restrictions. Moreover, we simulated the objective function value (φ in (10)) of OPF as the output of the grid model, which was iteratively triggered by the market transaction result of every timeslot, see Figure 17. In the figure, we noticed that the φ value of OPF is proportional to the total market trades (Aggregate Demand) of every timeslot, which shows the adaptation of the market clearing mechanism that takes the real-time power flow and transmission constraints into account.

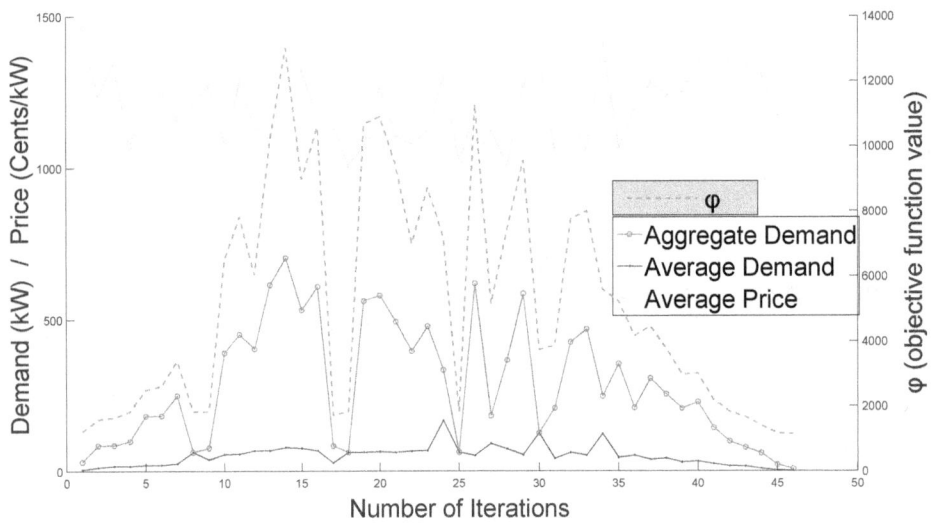

Figure 17.: Output of the power grid model with a market integration: demand is defined in unit *kW* and price is defined in unit *Cents/kW*; both demand and price are plotted with the same magnitude axis on the left

3.5 CONCLUSION

In this chapter, we presented the intra-minute market model and the two-layer grid model. The intra-minute market model is designed for a real-time price settlement as a balancing option that can be linked with the intraday or balancing market. The price settlement model refers to a dynamic market clearing among consumers, producers and network players, which determines grid-based nodal prices and further employes a local market price concept to stabilize the determined nodal prices. The two-layer grid model consists of a grid topology layer and a power flow layer. The grid topology layer describes the static part of the grid model, such as basic properties of the grid topology and components, while the power flow layer formulates the dynamic part of the grid model in terms of an OPF-based power flow model. Subsequently, we proposed a definition of the market-grid coupling and formalized a feedback control loop as the coupling approach. The nodal price is employed as the feedback signal for the proposed coupling model. After a short discussion of the coupling requirements from the control point of view, we demonstrated an initial test of this market-grid coupling model, which showed that a real-time adaptation of the power market clearing based on the grid optimization output is feasible. Since this chapter mainly provides a model description of the market and grid as well as a feedback control framework for constructing the market-grid coupling, the next chapter begins with a concrete control design based on the model description and control framework of this chapter, in order to investigate the mutual influence between the power market and grid with respect to both price and supply-demand-balance stability.

BIBLIOGRAPHY

[1] Ilge Akkaya, Yan Liu, and Edward A. Lee. *Cyber Physical Systems Approach to Smart Electric Power Grid*, chapter Modeling and Simulation of Network Aspects for Distributed Cyber-Physical Energy Systems, pages 1–23. Springer Berlin Heidelberg, Berlin, Heidelberg, 2015. ISBN 978-3-662-45928-7. doi: 10.1007/978-3-662-45928-7_1. URL http://dx.doi.org/10.1007/978-3-662-45928-7_1.

[2] F. Alvarado. The stability of power system markets. *Power Systems, IEEE Transactions on*, 14(2):505–511, May 1999. ISSN 0885-8950. doi: 10.1109/59.761873.

[3] G. Andersson. Modelling and analysis of electric power systems. Technical report, ETH Zurich, 2008.

[4] G. Andersson. Lecture notes of eeh - power systems laboratory. ETH Zurich, September 2012.

[5] M. Anghel, K.A. Werley, and A.E. Motter. Stochastic model for power grid dynamics. In *System Sciences, 2007. HICSS 2007. 40th Annual Hawaii International Conference on*, pages 113–113, Jan 2007. doi: 10.1109/HICSS.2007.500.

[6] R. Baldick. Electricity market equilibrium models: the effect of parametrization. *Power Systems, IEEE Transactions on*, 17(4):1170–1176, Nov 2002. ISSN 0885-8950. doi: 10.1109/TPWRS.2002.804956.

[7] Guillermo Bautista Alderete. *Alternate models to analyze market power and financial transmission rights in electricity markets*. PhD thesis, Waterloo, Ont., Canada, Canada, 2005. AAINR12143.

[8] BMWi. Zeitreihen zur entwicklung der erneuerbaren energien in deutschland. online, December 2015. URL http://www.erneuerbare-energien.de/EE/Navigation/DE/Service/Erneuerbare_Energien_in_Zahlen/Zeitreihen/zeitreihen.html. Based on Working Group on Renewable Energy-Statistics (AGEE-Stat).

[9] Frieder Borggrefe and Karsten Neuhoff. Balancing and intraday market design: Options for wind integration. 2011.

[10] Brandon Davito, Humayun Tai, and Robert Uhlaner. The smart grid and the promise of demand-side management. *McKinsey on Smart Grid*, pages 38–44, 2010.

[11] C.J. Day, B.F. Hobbs, and J.-S. Pang. Oligopolistic competition in power networks: a conjectured supply function approach. *Power Systems, IEEE Transactions on*, 17(3):597–607, Aug 2002. doi: 10.1109/TPWRS.2002.800900.

[12] Yi Ding, P. Nyeng, J. Ostergaard, Maj Dang Trong, S. Pineda, Koen Kok, G.B. Huitema, and O.S. Grande. Ecogrid eu - a large scale smart grids demonstration of real time market-based integration of numerous small der and dr. In *Innovative Smart Grid Technologies (ISGT Europe), 2012 3rd IEEE PES International Conference and Exhibition on*, pages 1–7, Oct 2012. doi: 10.1109/ISGTEurope.2012.6465895.

[13] Yi Ding, S. Pineda, P. Nyeng, J. Ostergaard, E.M. Larsen, and Qiuwei Wu. Real-time market concept architecture for ecogrid eu - a prototype for european smart grids. *Smart Grid, IEEE Transactions on*, 4(4):2006–2016, Dec 2013. ISSN 1949-3053. doi: 10.1109/TSG.2013.2258048.

[14] Yong Ding, M.A. Neumann, M. Budde, M. Beigl, P.G. Da Silva, and Lin Zhang. A control loop approach for integrating the future decentralized power markets and grids. In *Smart Grid Communications (SmartGridComm), 2013 IEEE International Conference on*, pages 588–593, Oct 2013. doi: 10.1109/SmartGridComm.2013.6688022.

[15] Yong Ding, Erwin Stamm, and Michael Beigl. Using model predictive control to interact distributed power markets and grids. In *in Proceedings of IEEE International Energy Conference (ENERGYCON 2016)*, Leuven, Belgium, April 2016. IEEE.

[16] ENTSO-E. Entso-e overview of internal electricity market-related project work, October 2014.

[17] Joseph H. Eto and Robert J. Thomas, editors. *Computational Needs for the Next Generation Electric Grid*, 2011.

[18] Debin Fang, Jingfang Wu, and Dawei Tang. A double auction model for competitive generators and large consumers considering power transmission cost. *International Journal of Electrical Power & Energy Systems*, 43(1):880 – 888, 2012. ISSN 0142-0615. doi: http://dx.doi.org/10.1016/j.ijepes.2012.05. 041. URL http://www.sciencedirect.com/science/article/pii/ S0142061512002311.

[19] Xi Fang, Satyajayant Misra, Guoliang Xue, and Dejun Yang. Smart grid - the new and improved power grid: A survey. *Communications Surveys Tutorials, IEEE*, 14(4):944–980, Fourth 2012. ISSN 1553-877X. doi: 10.1109/SURV.2011. 101911.00087.

[20] H. Farhangi. The path of the smart grid. *Power and Energy Magazine, IEEE*, 8 (1):18–28, January 2010.

[21] Martin Geidl and G. Andersson. Optimal power flow of multiple energy carriers. *Power Systems, IEEE Transactions on*, 22(1):145–155, Feb 2007. doi: 10.1109/TPWRS.2006.888988.

[22] M. Girvan and M. E. J. Newman. Community structure in social and biological networks. *Proceedings of the National Academy of Sciences*, 99(12):7821–7826, 2002. doi: 10.1073/pnas.122653799. URL http://www.pnas.org/content/99/12/7821.abstract.

[23] Per Goncalves Da Silva, Stamatis Karnouskos, and Dejan Ilic. Evaluation of the scalability of an energy market for smart grid neighbourhoods. In *IEEE 11th International Conference on Industrial Informatics (INDIN), Bochum, Germany*, July29–31 2013.

[24] Rasmus Halvgaard, John Bagterp Jørgensen, Niels Kjølstad Poulsen, and Henrik Madsen. *Model Predictive Control for Smart Energy Systems*. PhD thesis, Technical University of DenmarkDanmarks Tekniske Universitet, AdministrationAdministration, Office for Study Programmes and Student AffairsAfdelingen for Uddannelse og Studerende, 2014.

[25] Lion Hirth and Inka Ziegenhagen. Balancing power and variable renewables: Three links. *Renewable and Sustainable Energy Reviews*, 50:1035 – 1051, 2015. ISSN 1364-0321. doi: http://dx.doi.org/10.1016/j.rser.2015.04.180. URL http://www.sciencedirect.com/science/article/pii/S1364032115004530.

[26] B.F. Hobbs. Linear complementarity models of nash-cournot competition in bilateral and poolco power markets. *Power Systems, IEEE Transactions on*, 16 (2):194–202, 2001.

[27] B.F. Hobbs, C.B. Metzler, and J.-S. Pang. Strategic gaming analysis for electric power systems: an mpec approach. *Power Systems, IEEE Transactions on*, 15(2): 638–645, May 2000. doi: 10.1109/59.867153.

[28] Dejan Ilic, Per Goncalves Da Silva, Stamatis Karnouskos, and Martin Griesemer. An energy market for trading electricity in smart grid neighbourhoods. In *6th IEEE International Conference on Digital Ecosystem Technologies – Complex Environment Engineering (IEEE DEST-CEE), Campione d'Italia, Italy*, June 2012.

[29] Tooraj Jamasb and Michael Pollitt. Electricity market reform in the european union: review of progress toward liberalization & integration. *The Energy Journal*, pages 11–41, 2005.

[30] E. Kahn. Numerical techniques for analyzing market power in electricity. *Electricity Journal*, 11(6):34–43, 1998.

[31] A.R. Kian, J.B. Cruz, and R.J. Thomas. Bidding strategies in oligopolistic dynamic electricity double-sided auctions. *Power Systems, IEEE Transactions on*, 20(1):50–58, Feb 2005. ISSN 0885-8950. doi: 10.1109/TPWRS.2004.840413.

[32] Cen Li. Classifying imbalanced data using a bagging ensemble variation (BEV). In *Proceedings of the 45th annual southeast regional conference on - ACM-SE 45*, page 203, New York, New York, USA, 2007. ACM Press. ISBN 9781595936295. doi: 10.1145/1233341.1233378. URL http://portal.acm.org/citation.cfm?doid=1233341.1233378.

[33] Gong Li, Jing Shi, and Xiuli Qu. Modeling methods for genco bidding strategy optimization in the liberalized electricity spot market - a state-of-the-art review. *Energy*, 36(8):4686 – 4700, 2011. ISSN 0360-5442. doi: http://dx.doi.org/10.1016/j.energy.2011.06.015. URL http://www.sciencedirect.com/science/article/pii/S0360544211003926. {PRES} 2010.

[34] P. Mäntysaari. *EU Electricity Trade Law: The Legal Tools of Electricity Producers in the Internal Electricity Market*. Springer International Publishing, 2015. ISBN 9783319165134. URL https://books.google.de/books?id=krElCQAAQBAJ.

[35] Carolyn Metzler, Benjamin F. Hobbs, and Jong-Shi Pang. Nash-cournot equilibria in power markets on a linearized dc network with arbitrage: Formulations and properties. *Networks and Spatial Economics*, 3(2):123–150, 2003. ISSN 1572-9427. doi: 10.1023/A:1023907818360. URL http://dx.doi.org/10.1023/A:1023907818360.

[36] F. Milano. *Power Systems: Power System Modelling and Scripting*. Power systems. Springer Berlin Heidelberg, 2010. ISBN 9783642136696.

[37] Felix Muesgens, Axel Ockenfels, and Markus Peek. Economics and design of balancing power markets in germany. *International Journal of Electrical Power & Energy Systems*, 55:392 – 401, 2014. ISSN 0142-0615. doi: http://dx.doi.org/10.1016/j.ijepes.2013.09.020. URL http://www.sciencedirect.com/science/article/pii/S014206151300402X.

[38] D.E. Newman, B.A. Carreras, M. Kirchner, and I. Dobson. The impact of distributed generation on power transmission grid dynamics. In *System Sciences (HICSS), 2011 44th Hawaii International Conference on*, pages 1–8, Jan 2011. doi: 10.1109/HICSS.2011.414.

[39] J. Nicolaisen, V. Petrov, and L. Tesfatsion. Market power and efficiency in a computational electricity market with discriminatory double-auction pricing. *Evolutionary Computation, IEEE Transactions on*, 5(5):504–523, Oct 2001. ISSN 1089-778X. doi: 10.1109/4235.956714.

[40] Takashi Nishikawa and Adilson E Motter. Comparative analysis of existing models for power-grid synchronization. *New Journal of Physics*, 17(1):015012, 2015. URL http://stacks.iop.org/1367-2630/17/i=1/a=015012.

[41] Giuliano Andrea Pagani and Marco Aiello. The power grid as a complex network: A survey. *Physica A: Statistical Mechanics and its Applications*, 392 (11):2688 – 2700, 2013. ISSN 0378-4371. doi: http://dx.doi.org/10.1016/j.physa.2013.01.023. URL http://www.sciencedirect.com/science/article/pii/S0378437113000575.

[42] P. Palensky and D. Dietrich. Demand side management: Demand response, intelligent energy systems, and smart loads. *Industrial Informatics, IEEE Transactions on*, 7(3):381–388, Aug 2011. doi: 10.1109/TII.2011.2158841.

[43] M.K. Petersen, L.H. Hansen, J. Bendtsen, K. Edlund, and J. Stoustrup. Market integration of virtual power plants. In *Decision and Control (CDC), 2013 IEEE 52nd Annual Conference on*, pages 2319–2325, Dec 2013. doi: 10.1109/CDC.2013.6760227.

[44] M. Roozbehani, Munther Dahleh, and S. Mitter. Dynamic pricing and stabilization of supply and demand in modern electric power grids. In *Smart Grid Communications (SmartGridComm), 2010 First IEEE International Conference on*, pages 543–548, Oct 2010. doi: 10.1109/SMARTGRID.2010.5621994.

[45] M. Roozbehani, Munther Dahleh, and S. Mitter. On the stability of wholesale electricity markets under real-time pricing. In *Decision and Control (CDC), 2010 49th IEEE Conference on*, pages 1911–1918, Dec 2010. doi: 10.1109/CDC.2010.5718173.

[46] M. Roozbehani, M.A. Dahleh, and S.K. Mitter. Volatility of power grids under real-time pricing. *Power Systems, IEEE Transactions on*, 27(4):1926–1940, Nov 2012. doi: 10.1109/TPWRS.2012.2195037.

[47] C. Ruiz, A.J. Conejo, and Y. Smeers. Equilibria in an oligopolistic electricity pool with stepwise offer curves. *Power Systems, IEEE Transactions on*, 27(2): 752–761, May 2012. ISSN 0885-8950. doi: 10.1109/TPWRS.2011.2170439.

[48] Peter W Sauer and MA Pai. Power system dynamics and stability. *Urbana*, 1998.

[49] Manoj Sharma. Flow-based market coupling. Master's thesis, Delft University of Technology, August 2007.

[50] EPEX SPOT. Flow-based methodology for cwe market coupling successfully launched. PRESS RELEASE, May 2015.

[51] Steven Stoft. *Power System Economics: Designing Markets for Electricity*, volume 8. THE OXFORD INSTITUTE FOR ENERGY STUDIES, 2002.

[52] Tadahiro Taniguchi, Koki Kawasaki, Yoshiro Fukui, Tomohiro Takata, and Shiro Yano. Automated linear function submission-based double auction as bottom-up real-time pricing in a regional prosumers electricity network. *Energies*, 8(7):7381–7406, 2015.

[53] Tadahiro Taniguchi, Tomohiro Takata, Yoshiro Fukui, and Koki Kawasaki. Convergent double auction mechanism for a prosumers decentralized smart grid. *Energies*, 8(11):12342–12361, 2015.

[54] M. Ventosa, A. Baillo, A. Ramos, and M. Rivier. Electricity market modeling trends. *Energy Policy*, 33:897–913, 2005.

[55] Ovidiu Vermesan, Lars-Cyril Blystad, Roberto Zafalon, Alessandro Moscatelli, Kai Kriegel, Randolf Mock, Reiner John, Marco Ottella, and Pietro Perlo. *Advanced Microsystems for Automotive Applications 2011: Smart Systems for Electric, Safe and Networked Mobility*, chapter Internet of Energy – Connecting Energy Anywhere Anytime, pages 33–48. Springer Berlin Heidelberg, Berlin, Heidelberg, 2011. ISBN 978-3-642-21381-6. doi: 10.1007/978-3-642-21381-6_4. URL http://dx.doi.org/10.1007/978-3-642-21381-6_4.

[56] Rafal Weron. Electricity price forecasting: A review of the state-of-the-art with a look into the future. *International Journal of Forecasting*, 30(4):1030 – 1081, 2014. ISSN 0169-2070. doi: http://dx.doi.org/10.1016/j.ijforecast.2014.08.008. URL http://www.sciencedirect.com/science/article/pii/S0169207014001083.

[57] Liang Xie, Hsiao-Dong Chiang, and Shao-Hua Li. Locational marginal pricing under composite dynamic load models: Formulation and computation. In *Power and Energy Society General Meeting, 2010 IEEE*, pages 1–8. IEEE, 2010.

[58] R.D. Zimmerman, C.E. Murillo-Sanchez, and R.J. Thomas. Matpower: Steady-state operations, planning, and analysis tools for power systems research and education. *Power Systems, IEEE Transactions on*, 26(1):12–19, Feb 2011. doi: 10.1109/TPWRS.2010.2051168.

[59] Marco Zugno, Juan Miguel Morales, Pierre Pinson, and Henrik Madsen. A bilevel model for electricity retailers' participation in a demand response market environment. *Energy Economics*, 36:182 – 197, 2013. ISSN 0140-9883. doi: http://dx.doi.org/10.1016/j.eneco.2012.12.010. URL http://www.sciencedirect.com/science/article/pii/S0140988312003477.

4

SYSTEM MODELING OF A LOCAL MARKET-GRID CONTROL LOOP

4.1 ABSTRACT AND CONTEXT

In Chapter 3, the power market model and the power grid model that are employed as a modeling basis in this dissertation were presented. Based on both models, we defined a market-grid coupling concept and formalized a correspondent feedback control loop as the coupling approach. This chapter is concerned with a further investigation and analysis of this formalized market-grid coupling. First, we present a co-simulation framework to analyze the mutual influence of both market and grid in a simulation environment. Then, we extend the proposed feedback control system with a model predictive control (MPC) based dispatch control component, in order to investigate the interoperable controllability between the market and the grid within the market-grid coupling framework. As the first step towards a distributed market-grid coupling, we focus in this chapter on the simulation framework design and the control system design for coupling a single grid unit with a correspondent local market. After a short introduction of the MPC theory, we formulate the local market-grid coupling as a centralized MPC problem and present the mathematical formulation of this optimization problem. Based on different test scenarios with IEEE bus systems, we demonstrate simulation-based numerical results to show the capability of this MPC-based control concept for a local market-grid coupling with respect to the impact of an optimal dynamic dispatch on the market price stability. In the next chapter, the centralized MPC problem for a single grid unit will be extended towards a distributed MPC problem for distributed grid units. The content of this chapter consists of two parts. One is based on our co-simulation paper [6] and the other one is based on our MPC formulation paper [7], which are published at IEEE ENERGYCON 2016.

4.2 A CO-SIMULATION FRAMEWORK

In the previous chapter, we proposed a closed-loop feedback system that models the market-grid coupling by means of the LMP-based nodal price on the physical grid. In order to investigate the impact of dynamic power flows of the grid network on the electricity price and vice versa, we need either a real-world implementation or a simulation framework of the proposed market-grid coupling. Due to the research purpose as well as market and grid operational restrictions in the real world, it is not feasible to have a real-world test. Therefore, we opted for the simulation framework option to interconnect the market and the grid in the form of a co-simulation, in order to analyze the mutual influence on each other in a simulation environment.

In what follows, we first introduce the existing open source simulation solutions of the power market or the power grid, which are relevant for the market-grid coupling. Then, we present our co-simulation framework with design decisions. Afterwards, we demonstrate a performance evaluation of the market-grid co-simulation in terms of system utilization and scalability. Finally, we provide an analysis of a grid-driven pricing scheme that reflects real-time load situation and power supply availability. A further investigation and problem formulation of the MPC-based control system to model and simulate a market-driven power re-dispatch is the content of the next section.

4.2.1 *Relevant Simulation Tools*

The simulation tools that we focus on are all open source simulation solutions for the power market or the power grid. They enable the generation of the possibility that the power flow in the grid is generated based on the OPF formulation, in order to derive therefrom the LMP-based nodal prices. The solutions mentioned below are primarily designed for the US power grid that differs slightly from the German one. Compared to Germany, the power grid in the United States uses 60 Hz as grid frequency and other voltage ranges at low, medium and high voltage levels. However, these differences are not relevant to the principle of a power-flow study.

MATPOWER

MATPOWER is a package of MATLAB M-files developed by Zimmerman et al. [20] for solving power flow and optimal power flow problems. The MATPOWER tool alone can only present an one-shot simulation, and no successive simulation data is possible. According to authors [20], MATPOWER is for simple power flow calculations one of the fastest open source programs. Through its well-designed extendable implementation, it is suitable for simulation of large power grids as well as to be integrated into other programs as an OPF solver.

Power System Analysis Toolbox (PSAT)

PSAT is a MATLAB toolbox developed by Milano et al. [12] for power system analysis and simulation. The PSAT toolbox focuses mainly on analysis and simulation of small- or medium-sized power grids. For large power grids, it is not sufficiently fast compared to its competitors. PSAT supports the use of static and dynamic as well as user-defined models. However, no temporal schedule function is available for simulation of load profiles. Otherwise, the power system analysis in PSAT features not only power flow study, but also almost all other power system analysis problems, such as small signal stability analysis, fault analysis, etc.

GridLAB-D

GridLAB-D [4] is an open source modeling and simulation tool developed by the US Department of Energy at Pacific Northwest National Laboratory (PNNL). It focuses on simulation and analysis for distribution power grid. Besides network modules within the GridLAB-D tool, which are responsible for the feed-in, there are various types of load models that can be used to create typical load profiles of different consumers either affected by climate data of a realistic environment or controlled by temporal schedules or prerecorded load curves. In addition, GridLAB-D has a market module for power trading, in which the classic double auction mechanism is implemented, but a calculation of nodal prices based on the OPF is not yet possible.

AMES Wholesale Power Market Test Bed

AMES is a power market test bed [11] developed at Iowa State University for the wholesale market design proposed by the U.S. Federal Energy Regulatory Commission. The test bed is an agent-based framework that can be used to simulate market behaviors of producers (GenCos: Generation Companies) and consumers (LSEs: Load Serving Entities). The Independent System Operator (ISO) agent is responsible for the market clearing, by determining nodal prices for the day-ahead market based on received load profiles of LSEs and generation offers of GenCos. The AMES test bed provides the feature of calculating LMP-based nodal prices based upon the OPF formulation of a physical grid.

GridSpice

GridSpice is a Python-based co-simulation framework developed by Anderson et al. [2], which integrates MATPOWER and GridLAB-D as simulation programs. GridSpice aims at modeling the interactions between all components of a power grid. Its current implementation focuses on distribution simulations along with one-shot optimal power flow and demand response mechanisms. However, a support for market integration is yet to be provided as future work. Unfortunately, the only source code of 2012, which is accessible in a public repository, seems to not function any more; and the bug tracker of the project as well as the discussion groups appear no longer to be maintained.

AMES/GridLAB-D Test Bed

Aliprantis et al. [1] introduced a test bed for modeling the integration of retail and wholesale markets operating over physical grid dynamics and constraints. This test bed refers to a co-simulation platform that integrates AMES and GridLAB-D as simulation tools, where AMES generates a real-time price and GridLAB-D submits subsequently the load data. The project of AMES/GridLAB-D Test Bed focuses on investigation of price responsiveness to power system operations. Authors concern themselves with research questions, such as what kind of distributed energy resources and pricing mechanisms provide the most social efficiency, or whether a change from the transmission network to microgrids with respect to functionality and efficiency pays off. At the time of preparation of our simulation framework development, to the best of our knowledge, the described Test Bed exists only as a concept. Our co-simulation framework in terms of a market-grid coupling tries to extend the concept of AMES/GridLAB-D Test Bed towards with an enhanced interoperability. That means that we focus on a simulation framework, which provides not only features for analyzing grid-driven pricing schemes but also possibilities for constructing feedback control concepts to model and simulate a market-driven power re-dispatch.

4.2.2 *Framework Design and Implementation*

In this subsection, we describe the design and implementation of our co-simulation framework for the proposed market-grid coupling. The framework enables an identification of nodal prices based on synthesized load profiles and can feed back a price-dependent load change into grid. Moreover, we discuss the choice of components and procedures as well as architecture decisions.

For reasons of portability and flexibility regarding extensions, we chose to develop the framework in the dynamic Python programming language. Python is distinguished by the fact that its programs are executable without prior compiling from a standing start and achieves a very large coverage in terms of operating systems and platforms. The main components of the framework are designed as *Controller* and *Worker*. The *Controller* is responsible for the simulation control and modeling the power market, while the *Worker* functions as load profile simulator regarding dynamics and constraints of a power grid. Furthermore, a configuration scheme is developed, which minimizes redundancies and contains the necessary configuration parameters. An overview of the framework architecture is depicted in Figure 18. As shown in the figure, the main information that are transferred between the simulation platforms refers to *price* and *load*. Details of each system component are given in the following paragraphs.

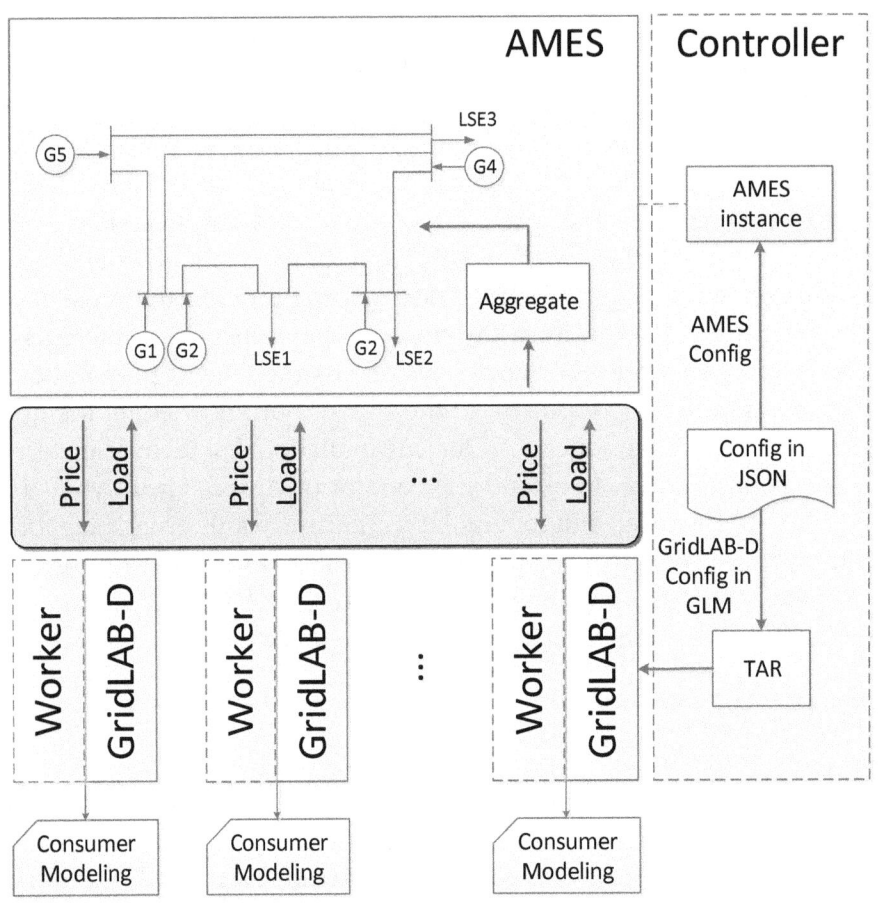

Figure 18.: An overview of the framework architecture

Selection of Simulation Tools

Based on the analysis in Section 4.2.1, GridLAB-D and AMES are proven to be suitable components of the proposed co-simulation for a market-grid coupling, which meets our requirements. Within the framework, GridLAB-D is responsible for modeling different aggregated load profiles. These aggregated loads consist of individual consumption loads of such devices as dishwashers or microwaves, as well as whole houses or entire settlements, which model appropriate power consumption behaviors. Within GridLAB-D, power loads are modeled in a GLM file that specifies different power flow and load models with three main declaration blocks, i.e. *clock*, *module* and *object* [4]. To increase the model reliability, it is also possible to integrate climate models and thus analyze the loads (and indirectly also the prices) with different climatic conditions. In order to transform the day-ahead market simulation of AMES to a real-time market (pricing), it is required that a representation of load profiles is needed on a time scale of exactly 24 hours or multiples of 24 hours. This is the only requirement that the framework imposes on GridLAB-D. Apart from this, the loads can be modeled with GridLAB-D freely. Our preliminary test showed that the integrated multi-threading of GridLAB-D is implemented insufficiently and it causes heavy performance losses due to existing locking mechanisms. Furthermore, the performance losses are increased by the multi-threaded communication effort in particular in short-term calculations. This deficiency will be compensated to some extent by the framework implementation.

As mentioned in Section 4.2.1, AMES represents loads and generators as so-called LSEs and GenCos, respectively. A GenCo aims here at its maximum profit, while a LSE tries to achieve the lowest possible cost. The LSEs and GenCos are located on buses, which are connected to each other via branches. Minimizing the LSEs' costs in consideration of the capacity and impedance information of the grid, the power flows in AMES are formed. AMES is based on Java and uses Swing as graphics library and API. For this reason, modifications on AMES and its GUI have to be made, so that the configuration of a power grid does not need to be transferred through the GUI and AMES can be then used in conjunction with the framework. These changes are also required for the transformation of a real-time market, so that an AMES market clearing can be conducted in real time once the information of power loads is transferred from GridLAB-D.

Communication Architecture

Since GridLAB-D is currently only executable under Linux with limitations and scalability requirements have to be achieved, the framework components are deployed on different computing units at the same time. This presupposes the application of a communication architecture. Due to the separate simulation components, no inter-process communication mechanisms, such as shared memory or message passing, can be used. Thus, data exchange on the network level is required. For a network-technical connection of two simulation components, we implement a client-server model that offers an easy way to communicate between processes on different systems.

Controller

The *Controller* is the key of the whole framework and is responsible for the control of the co-simulation. The *Controller* receives a configuration file and interprets it, afterwards controls further program runs based on the configuration file. In case a feedback controller for a price-dependent load change is integrated at the defined interface, the *Controller* will determine the load change according to the calculated market price.

Since GridLAB-D and AMES use two different and incompatible configuration formats, the *Controller* builds unified configuration files for both components and can read and further process the results from the program calculation. As part of this task, `buildconfig()` takes over a parameterization of AMES for calculating nodal prices. To simplify and minimize redundancies, a syntax based on JSON (JavaScript Object Notation) for the configuration file is created. A simplified class diagram of the framework consisting of the *Controller* class, the *Worker* class and the *AMESconfig* class can be seen in Figure 19.

A simulation process begins as shown in the sequence diagram in Figure 20 with an initialization of the *Controller* class. It receives as parameter the file name of a JSON-formatted configuration file. The *Controller* class checks then the existence of JAVA, imports via the method `importConfig()` the configuration and creates in the method `createAMESConfig()` the global variable `AMESconfig`, in order to collect the configuration data. Via the method `execute()`, the GridLAB-D configuration is packed in an archive and uploaded to the *Worker* server that is defined in the configuration file. The further detailed procedure on the *Worker* is detailed in 4.2.2. While the *Worker* calculates the load profiles, the *Controller* operates busy-waiting and downloads the calculated load profiles after the call of the *Worker*. Assuming that the *Worker* in this co-simulation model represents the limited resource, we consider the busy-waiting in this context to be uncritical. After downloading, the individual load profiles are allocated to the correspondent LSE based on its name and the associated load values are stored via all `add()` methods in the *AMESconfig* class. In case GridLAB-D generates load profiles for more than 24 hours, the *Controller* will produce nodal prices per 24 hours, respectively. In case the modeling approach of a virtual power plant is to be simulated,

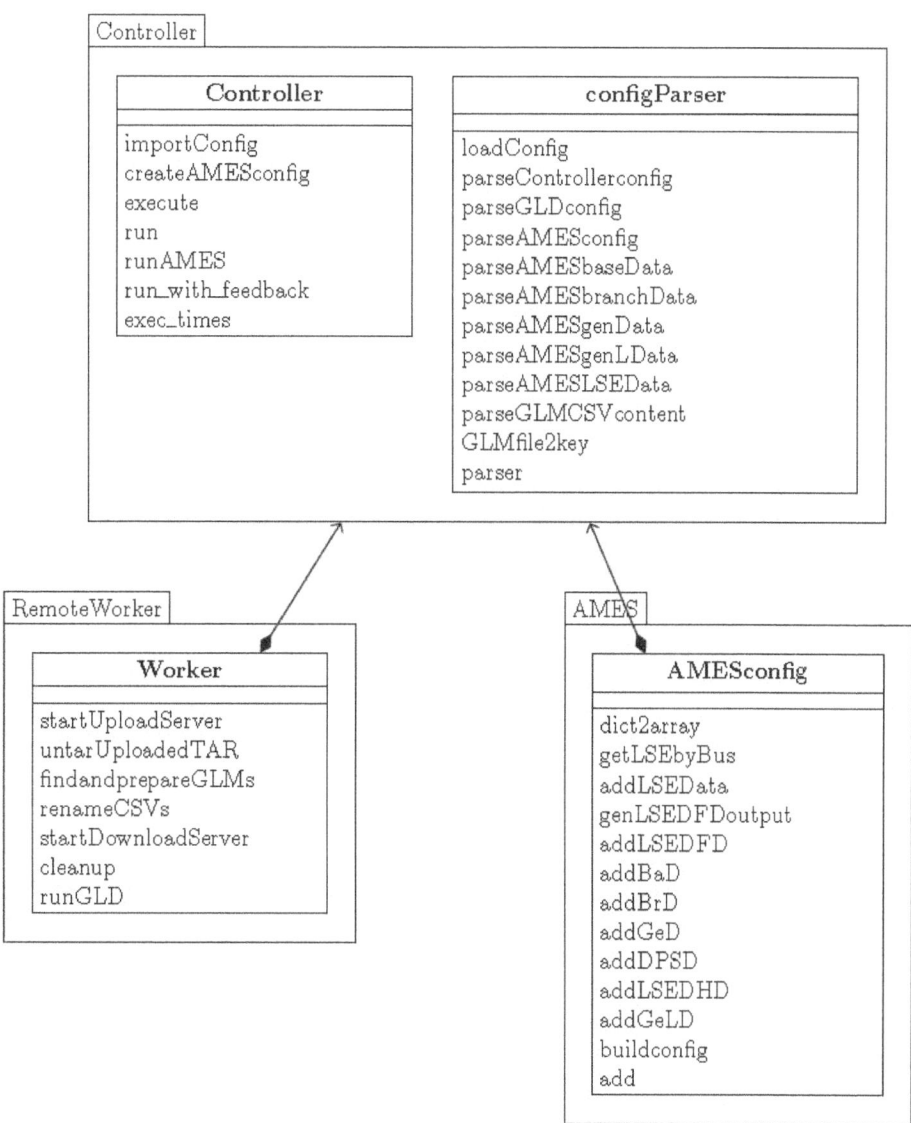

Figure 19.: A simplified class diagram of the framework

individual loads can be combined accordingly. If errors occur on the *Worker*, the *Controller* receives them in an *error.txt* file and processes them further.

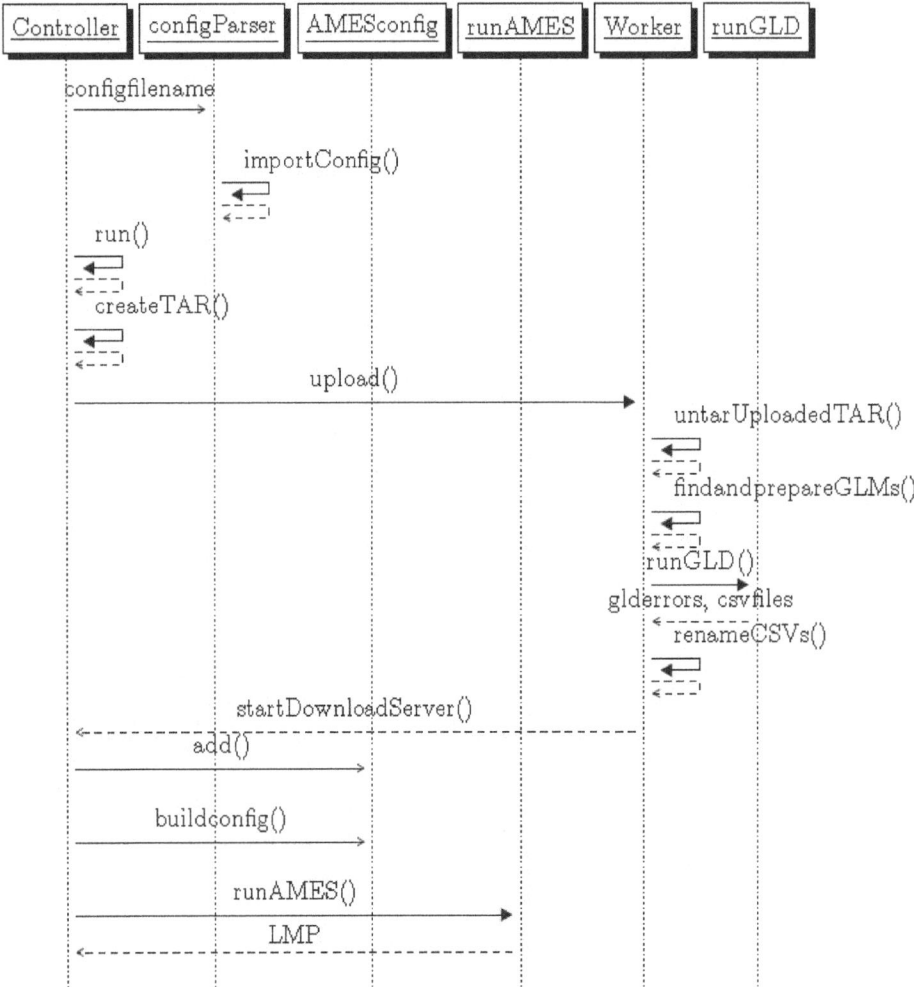

Figure 20.: The sequence diagram of the framework classes

For further integration of other tools, e.g. a feedback coupling program of price changes on a load change, it is possible via the `run()` method (see Figure 19), to connect and execute feedback controllers as proposed.

Communication

The data transfer in the communication network between the *Controller* and the *Worker* concerns mainly the configuration data of GridLAB-D, which are needed for its execution. In order to model a distributed system, we adapt the HTTP protocol to an appropriate communication protocol for the communication between the *Controller* and the *Worker*. An overview of the applied HTTP status codes for our framework is shown in Table 2.

HTTP status code	Meaning
201	File created successfully
400	General error
420	Archive faulty

Table 2.: HTTP status codes used in the framework

As mentioned before, the *Controller* parses the GridLAB-D configuration files (GLM files) rather in an archive file than raw GLM files. In comparison to embedding a GLM file as a string in an XML file, this design decision is superior due to less overhead. The *Controller* receives the information about, which GLM files are needed for the current calculation, via the central configuration file that specifies, which GridLAB-D configuration is allocated to which LSE. The GLM files are then compressed in a TAR archive and passed to the *Worker*. After receiving the TAR file, the *Controller* checks the file's validity. If the TAR file is correct, the *Worker* confirms the upload process with an HTTP status code 201 ("Item created") as success. If the TAR file can not be unzipped, the *Worker* reports the HTTP status code 420. If creating the TAR file is unsuccessful on the part of the *Worker*, the *Controller* receives the status code 400 as feedback. This may be caused by incorrect write permission or insufficient memory space. Finally, the *Worker* terminates the HTTP connection.

Worker

The *Worker* is responsible for modeling the correspondent load profiles on behalf of the *Controller* and provides the load information to the *Controller* for a further processing. In that case, the *Worker* execute the GridLAB-D instances to generate load profiles. For an optimal system utilization, the calculation of load profiles will take place in individual threads concurrently, so that each CPU core can be used if necessary.

For this purpose, the *Worker* receives individual GridLAB-D configuration files from the *Controller* via a HTTP connection. These files are unzipped by the *Worker* and the name of the result file will be stored in the Dictionary `lsecsvMapper`, so that the results of individual GridLAB-D calculations can also be allocated to the individual LSEs of AMES after the execution. After running GridLAB-D instances, the result files that are presented in CSV format, will be renamed to the LSE filenames and provided to the *Controller* in a TAR archive via HTTP.

Configuration File

As mentioned before, the *Controller* receives a set of configuration parameters for the underlying simulation model of the grid and the market via a configuration file. This configuration file minimizes the redundancies, improves the usability and simplifies the configuration of different tools through information hiding. We decide to employ JSON format for the configuration file, so that the individual objects of AMES and GridLAB-D are detached from their own systems and thus redundancies in the mapping of loads in GridLAB-D to a LSE in AMES are prevented.

The configuration file is flexibly expandable and currently consists of the following objects: 1) *Controllerconfig*; 2) *GLDconfig*; 3) *AMESconfig*; 4) *BaseData*; 5) *BranchData*; 6) *GenData*; 7) *LSEData*. Logically related elements are summarized here under the same objects. If data is required for both GridLAB-D and AMES, only one object for the data is needed. In the *Controllerconfig* object, it is determined which *Worker* is to be responded by the *Controller*.

4.2.3   *Performance Evaluation*

To test the framework performance, we deployed the co-simulation in a KVM (Kernel-based Virtual Machine) with 32 cores (each 2.3GHz) and 32GB of usable memory available. The 32 cores are exclusively allocated for this virtual machine. The operating system is 64-bit Windows 7 Enterprise. GridLAB-D is tested in the version 3.0.0-4524 (Hassayampa) and AMES in the version 2.06 involving the use of Sun Java 1.8.0_25. On this test machine, we conducted tests in terms of the system utilization and the run-time linearity with different load model sizes. In order to reduce statistical errors, the result in each test case is presented as a statistical average based on 100 simulation runs.

GridLAB-D

Due to the real-time requirement, the test on GridLAB-D refers to run time and scalability with respect to the number of simulated houses located at a distribution feeder that represents aggregated load profiles. A pre-determined number of houses with static loads and static schedules are tested, respectively. In our test house, we simulate an aggregate load by a water heater that heats a water tank (50 gallons capacity) with a heating power of 4.5kW.

For three different simulation times, i.e. 1 day, 10 days and 30 days, we observed that the memory is only filled at the beginning of the run and remains then constant throughout the entire run time. For each simulation time, the RAM usage remains fairly constant for a simulation of up to 100 houses. Then, the memory usage increases for more than 100 houses both in case with static loads and static schedules. Unfortunately, no convergence with regard to memory usage could be specified, since the memory usage is highly dependent on the used GridLAB-D modules and the complexity of the built load model scope.

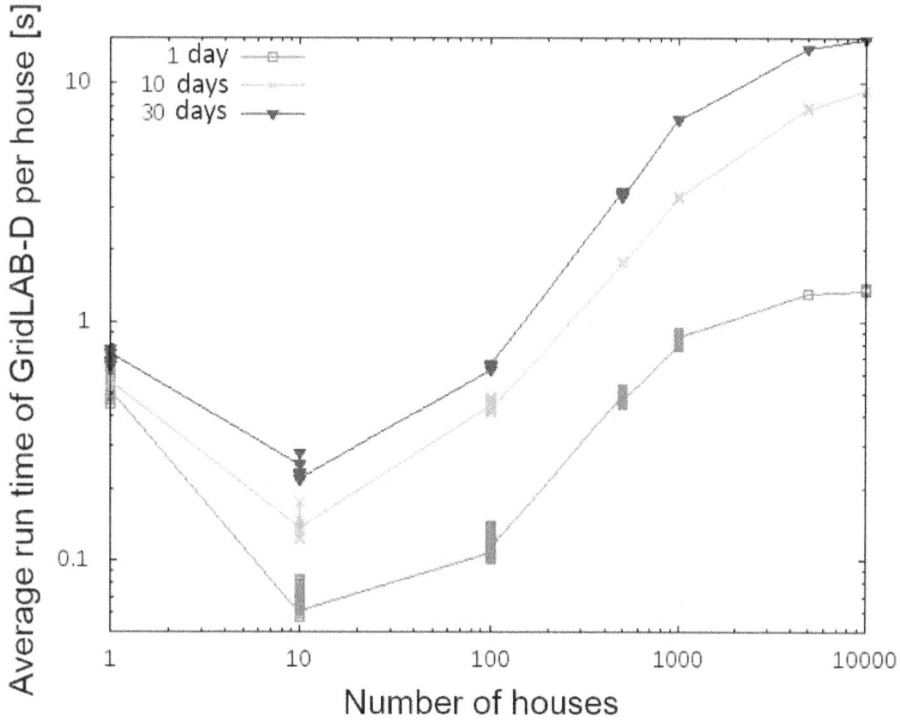

Figure 21.: Average runtime of GridLAB-D for simulating a house in conjunction with three different simulation times: 1 day, 10 days and 30 days

Regarding run time, a convergence in terms of average run time of GridLAB-D per house can be seen in Figure 21 for all simulation runs in conjunction with three simulation times. With an increasing number of houses, the probability for a different consumer behavior of the underlying stochastic model diminishes. Therefore, the figure shows a symbiotic behavior. Moreover, the run time does not behave exponentially but in proportion to the increasing number of houses. A further investigation regarding a more complex and realistic representation of the aggregate load is our future work.

AMES

In Table 3, an overview of the run time results of AMES is summarized according to the the number of simulated buses. However, for simplicity in AMES, the power flow study for an AC power network is approximated as a DC optimal power flow problem. AMES solves the optimization problem with an open-source Java solver, i.e. *QuadProgJ*, which is developed for strictly convex quadratic programming problems. The *QuadProgJ* solver seems ideal for an application of the DC optimal power flow calculation, which promises a high convergence [18] and solves extensive problems in a reasonable time.

Table 3.: Average run time of AMES depending on the number of simulated buses

Number of buses	AMES run time [s]
5	3
10	4
30	5
50	13
100	45

4.2.4 *Use Case: Nodal Price Analysis*

The current development of this framework enables mainly an open-loop analysis regarding the market price regulation, which means that 1) first, a load balancing is established through the OPF model, 2) then, based on the balancing results, the electricity power is traded accordingly. Therefore, in this subsection, we demonstrate the developed co-simulation framework for analyzing the influence of load changes on the market price. As a possible grid scenario, we modeled a supply congestion in the transmission system to establish to what extent an undersized transmission line with respect to transmission capacity has an influence on the nodal price and whether bypassing the supply congestion is reflected in the nodal price.

To simulate the grid scenario with overload or congestion, a grid topology is proposed as depicted in Figure 22a, which represents an exemplary power supply situation in Germany or a power transmission example across the Alps. In both cases, a lot of electrical energy is produced in the north and consumed in the south. In this scenario, the electricity has to flow over a constrained grid infrastructure, i.e. limited transmission capacity, and thereby causing a congestion situation. The nodes A and B are demonstrated representatively as a complex regional power grid, respectively, which have only little load demand on their own regional grid

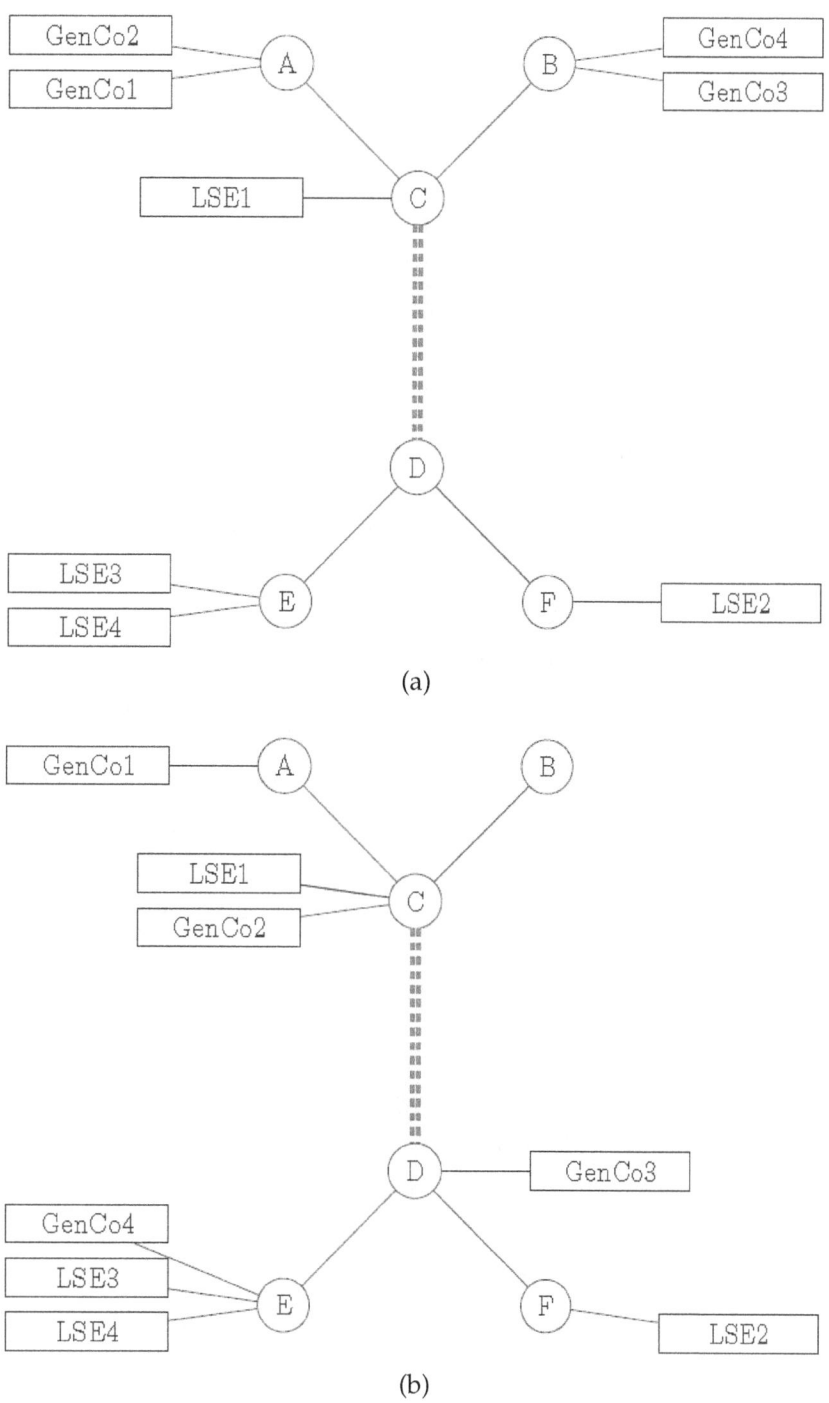

Figure 22.: A grid scenario with overload or congestion - (a) Overview of the proposed grid scenario with generation configuration case 1; (b) Overview of the proposed grid scenario with generation configuration case 2

and can therefore feed a large amount of electricity into the transmission grid. In contrast, the nodes E and F are representatives of regional grids, in which consumers are dominant, thereby taking much electricity necessarily from the transmission grid. The congestion or overload is symbolized in this transmission grid through the highlighted connection line between the nodes C and D. After a reallocation of the generation capacity from the nodes A and B to the nodes C, D and E as shown in Figure 22b, this provides another test case for a grid relief.

The grid parameters in terms of generation capacity and load that are used for the following analysis are summarized in Table 4. The same as the following transmission parameters, they all refer to synthetic grid data that don't comply with any grid model in the real world. However, it is important to mention that we can use this set of data to fairly reflect the overload situation of the grid. The percent values in the load field reflect the utilization of the transmission path between C and D. The parameters of transmission capacity and transmission impedance for the grid simulation are shown in Table 5.

Table 4.: Grid parameters of the proposed grid scenario in Figure 22

Node	Load [MW]					Generation [MW]	
	0%	25%	50%	75%	100%	Case 1	Case 2
A	-	-	-	-	-	500 + 500	500
B	-	-	-	-	-	500 + 500	-
C	0	37.5	75	112.5	150	-	500
D	-	-	-	-	-	-	500
E	0	137.5	275	412.5	550	-	500
F	0	125	250	375	500	-	-

Table 5.: Transmission parameters of the proposed grid scenario in Figure 22

From Node	To Node	Capacity [MW]	Impedance [MΩ/km]
A	C	2000	0.0060
B	C	2000	0.0060
C	D	1050	0.0060
D	E	3000	0.0060
D	F	3000	0.0060

To determine the nodal prices, two different pricing mechanisms, i.e. LMP and DA (Double Auction) [5] are implemented on each node with the same objective that generators gain the same profit regardless of pricing mechanisms. Table 6 shows a comparison of nodal prices with both LMP and DA applied in the test case 1, i.e. the grid scenario in Figure 22a. LMP behaves proportionally to the generation and transmission utilization of the congestion location, since calculation of nodal prices in accordance with LMP requires an overview of the power flow information of the entire grid. The nodes A and B benefit from their own high generation capacity, while at nodes E and F due to the congestion in the transmission path between C and D the congestion component of LMP determines nodal prices for a heavy overload. Furthermore, all nodes receive the same nodal price under zero congestion.

Table 6.: A comparison of the calculated nodal prices based on both LMP and DA

Node	Load-related Price									
	0%		25%		50%		75%		100%	
	LMP	DA	LMP	DA	LMP	DA	LMP	DA	LMP	DA
A	8	9	9	9	10	10	10	9	11	9
B	8	10	9	10	10	9	10	9	11	9
C	8	50	19	49	31	50	42	49	54	49
D	8	99	38	103	68	104	98	100	129	102
E	8	149	48	150	88	147	128	150	168	148
F	8	150	47	150	86	150	125	150	164	150

As seen in Table 6, the different congestion situations do not affect nodal prices in accordance with DA. This is due to the fact that in the DA pricing mechanism alone, there is no information available about the congestion situation.

In order to compare both pricing mechanisms from another perspective, we consider the correlation between the corresponding nodal price and the same load situation. In Figure 23, Pearson's correlation coefficients are depicted for both mechanisms, where x-axis denotes each individual simulation run (in total 100 runs). The DA-based price exhibits a correlation to the load demand in the range between -0.5 and 0.5, while the correlation of the LMP-based price to the load demand is in the range between 0.5 and 1. Furthermore, the significance of the correlation coefficients is verified based on the p-value. In comparison to the DA-based price, the LMP-based price shows a significant correlation to the load demand with a p-value less than 0.05.

Finally, Table 7 shows nodal prices in accordance with LMP for the test case 2, i.e. the grid scenario in Figure 22b. As mentioned before, the test case 2 refers to a reallocation of the generation capacity on the grid nodes, in order to achieve a grid relief from the test case 1. As seen in the table, the generator reallocation diminishes the congestion situation on the transmission path between C and D,

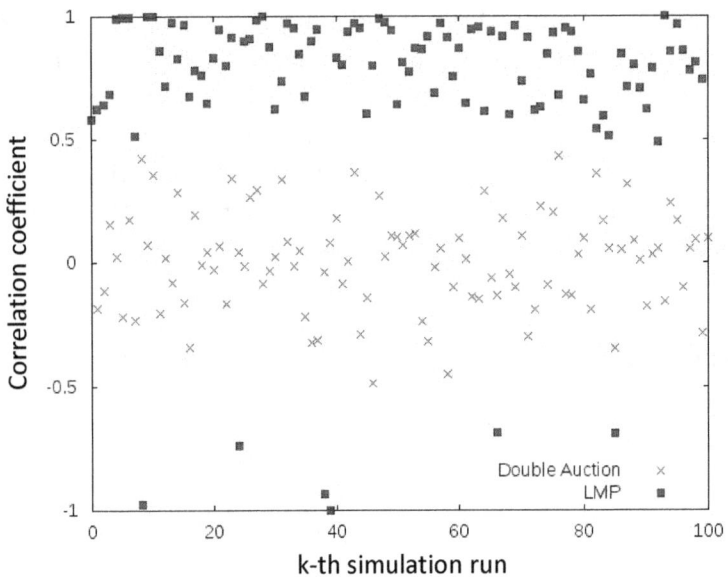

Figure 23.: Pearson's correlation coefficients for LMP and Double Auction

which makes the congestion component not any more as a price setter. Now the load demand on the individual nodes determine the price.

Table 7.: Calculated nodal prices based on LMP for the grid scenario in Figure 22b

Node	Load-related Price				
	0%	25%	50%	75%	100%
A	8	8	8	9	9
B	8	8	8	9	9
C	8	9	11	12	16
D	8	9	11	12	20
E	8	9	11	12	23
F	8	18	28	39	55

As shown in Table 6 and Table 7, the simulation results indicate that a grid-driven pricing mechanism, such as LMP, reflects well the real-time situation of the grid load. This confirms that our co-simulation framework can be used to analyze the mutual influence of market and grid. Moreover, this framework also demonstrates the ability for the optimization of power plant sites or as a grid expansion simulator.

4.2.5 *Summary*

In this section, we proposed a co-simulation framework that realizes a market-grid coupling and enables an analysis based on it for studying the load influence on the market price in consideration of physical grid dynamics and constraints. After introduction and comparison of the existing open source simulation tools, we decided to implement the co-simulation framework with GridLAB-D and AMES as grid and market simulation component. With the design decision of a client-server model, we described the implementation and the main functionality of each framework component including *Controller*, *Worker*, *Communication* and *Configuration*. In order to evaluate the framework, we conducted not only performance tests, but also a nodal price analysis for a defined grid scenario with transmission congestion. The results of system utilization showed that the memory usage is highly dependent on the complexity of the simulated grid model, though the convergent run time indicates a good scalability of the simulation framework. The nodal price analysis based on the framework could verify that the nodal price calculated based on LMP reflects well the real-time situation of the power grid. This concludes that our co-simulation framework can use a grid-driven pricing mechanism to bring the power market much closer to the power grid, in order to realize a simulation environment for the market-grid coupling. In this co-simulation framework, the proposed market-grid coupling has been only analyzed in terms of the grid impact on the market price, investigation of the reverse aspect, i.e. market price influence on grid load dispatch is the main content of the next section, which focuses on the design of a MPC-based feedback control concept for a grid re-dispatch driven by the power market.

4.3 A MPC-BASED FEEDBACK CONTROL SYSTEM

The previous section presented a co-simulation framework for the analysis of the proposed market-grid coupling with a focus on the relationship between the market price and the grid situation. Since the analysis refers mainly to the investigation of a grid-driven pricing mechanism, from a control theoretical point of view, it addresses an open-loop question to the market price control. However, the other aspect, i.e. a grid load re-dispatch based on the market price, is not considered yet. In order to guarantee the stability of the market-grid coupling, which is defined in the Introduction chapter and refers to the ability of a closed-loop feedback control system to maintain the steady-state operation of both the power market and the power grid, we need a closed-loop approach to address not only an optimal market price control but also a stable optimal dispatch.

One promising approach is the MPC-based (model predictive control) control system. As shown in the Related Work chapter, due to its inherent properties of both control theory and optimization theory, the MPC technique has been proposed by numerous recent research work as a closed-loop control mechanism for solving the ODD (Optimal Dynamic Dispatch) problems. In conventional control systems, e.g. the control loop for our proposed market-grid coupling defined in Chapter 3, the objective is to minimize the unconstrained error between the reference value and the actual measurement of the system output. However, a MPC-based approach for the ODD problems, known as economic MPC [14, 3], takes into account an economic term in the objective function of the control system. In the economic MPC case, the control task refers to an economic performance optimization (either a cost minimization or a profit maximization) rather than a setpoint tracking.

The objective of the closed-loop control system in this dissertation considers not only a power dispatch regulation but also a market price control. A MPC-based approach such as the economic MPC variant, is qualified for maintaining the supply-demand balance of the grid and in the meantime maximizing the profits (welfare) of all market participants. Both key features of the MPC concept, namely a dynamic optimization in a receding horizon manner and an optimization-based reactive control, are exploited in the design of our feedback control system, so that the two parts of the proposed control objective can be combined in one control loop, i.e. feedforward price control based on grid performance predictions over a predefined time horizon and feedback dispatch control using price measurements as input. In other words, our work extends the economic MPC problem for an optimal power dispatch with a market module, in order to study the impact of optimal dynamic dispatch on the market price stability as well as the importance of dynamic pricing for achieving a steady-state stability of the grid. In the following sub-sections, we first briefly introduce the control principle of a MPC, and then, present the system design of a MPC-based control loop for both grid-based price control and market-based power dispatch.

4.3.1 *Model Predictive Control*

Model Predictive Control (MPC) is an advanced control technique that formulates the process control task in an optimization problem. In comparison with the conventional control approaches, such as PID (proportional-integral-derivative) and LQR (linear-quadratic regulator) controllers, MPC features predictive capability in a finite time horizon that is commonly defined as the receding prediction horizon. Both MPC and LQR are approaches concerning optimal control theory. Besides the predictive capability, another main difference between the two approaches is that the MPC control law is normally formulated as a constrained optimization problem, while the LQR is unconstrained. Following the description of the MPC theory in the survey paper of Garcia et al. [9] and the book of Rawlings et al. [13], we present briefly the concept and principle of MPC in this sub-section.

MPC can be mathematically formulated as an iterative control optimization over a future time horizon of T steps. As an example, it is assumed that the system plant can be described by a state-space representation with the state variable $x(t)$, the controlled variable $y(t)$ (also known as system output variable) and the control variable $u(t)$. Then, with a predefined cost function J over the prediction horizon T, which represents the optimization objective in dependence on these system variables, a MPC-based optimal control problem can be formulated as follows.

$$\min_{u} \quad J(r, y, u) = \sum_{t=k}^{k+T-1} j((y(t) - r(t)), u(t)), \quad T \in \mathbb{N} \tag{21}$$

$$\text{subject to} \quad x(t+1) = f(x(t), u(t))$$
$$y(t) = g(x(t), u(t))$$
$$x_{min} \leq x(t) \leq x_{max}$$
$$u_{min} \leq u(t) \leq u_{max}$$
$$y_{min} \leq y(t) \leq y_{max}$$

where k stands for the current time step and $j(\bullet)$ represents the cost function at each time step, which takes two dependent variables into account, i.e. the deviation between system output $y(t)$ and the reference value $r(t)$, and the control input $u(t)$. $f(\bullet)$ and $g(\bullet)$ denote the common two system equations of a state-space representation. All three functions $j(\bullet)$, $f(\bullet)$ and $g(\bullet)$ can be either linear or non-linear. $x_{min/max}$, $u_{min/max}$ and $y_{min/max}$ describe the constraints on $x(t)$, $u(t)$ and $y(t)$, respectively. The objective in this formulation is related to minimizing costs.

Based on the above formulation, the principle of MPC can be illustrated in Figure 24. A reference trajectory $r(t)$ is given and defined as the set point, which is pursued by the system variable $y(t)$. At the current time $t = k$ that refers to the upper part of Figure 24, the current system output is measured and both past system putout and control input before k are available. Using available historical data and the current measurement, a state estimation procedure for the system output $y(t)$ is conducted based on the provided model of the state-space representation, starting at the current time k, over the prediction horizon T. Then, the cost function J is minimized for the entire prediction horizon via a numerical minimization algorithm, in order to determine a cost-minimizing control strategy in terms of a time series of the control input $u(t)$ from k to $k+T-1$. This time series is denoted as predicted control in Figure 24. Only the first optimal move of the time series is applied in the system plant. Afterwards, the behavior of the system plant is updated when a new measurement of $y(t)$ at the time $t = k + 1$ is available. The real-time optimization procedure is repeated for the new current time $k + 1$ with the forwards shifted prediction horizon, as shown in the bottom part of Figure 24. Then, both $y(t)$ and $u(t)$ are computed again, starting at the current time $k + 1$, over the prediction horizon T. This repeated optimization procedure with a fixed prediction horizon length (is often referred to in process control as a receding horizon control strategy) provides the property of a closed-loop feedback and can be used to compensate inaccuracies originating from model uncertainties.

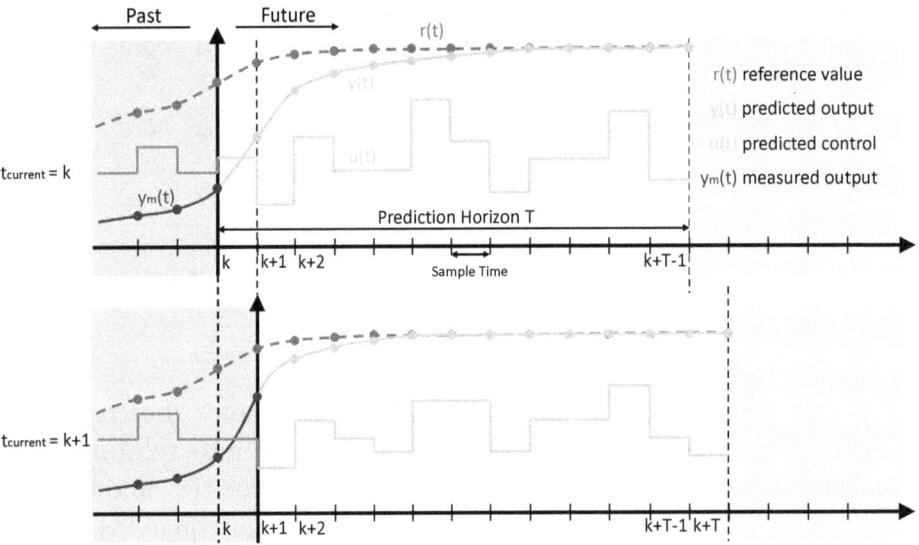

Figure 24.: A schematic representation of the receding horizon control principle of MPC

4.3.2 Control System Design

We build on our previous work [5] and the coupling idea of Roozbehani et al. [17], using LMP-based nodal prices to bridge the gap between the market and the grid, and then taking price as a feedback signal to control the dispatch schedule of generations units that preserves OPF. The control framework model adopted in this dissertation differs from the one in the work of Roozbehani et al. [17] that focused mainly on the effects of real-time retail pricing on the load volatility of a grid. In this work, we aim to achieve not only an optimal market price control but also a stable optimal dispatch, which refers to a study of mutual effects between the market and the grid. In addition, we focus on cooperation aspects among distributed grid units in terms of information exchange, in order to provide a collaborative capability for suppressing constraints' violation. In this chapter, we focus on the control system modeling for a local grid unit. The collaboration concept regarding a distributed MPC strategy for distributed grid units will be introduced in the next chapter.

In Chapter 3, we presented a closed-loop feedback system that models the coupling between a local power market and a local power grid by means of the LMP-based nodal price. As depicted in Figure 16, the proposed market-grid coupling is similar as the work of Roozbehani et al. [17] and has been mainly analyzed with the grid as the system plant and the market as a controller. In order to study mutual effects between the market and the grid, the existing closed-loop feedback system is extended by optimization-based control components towards a closed-loop feedback control system, see Figure 25.

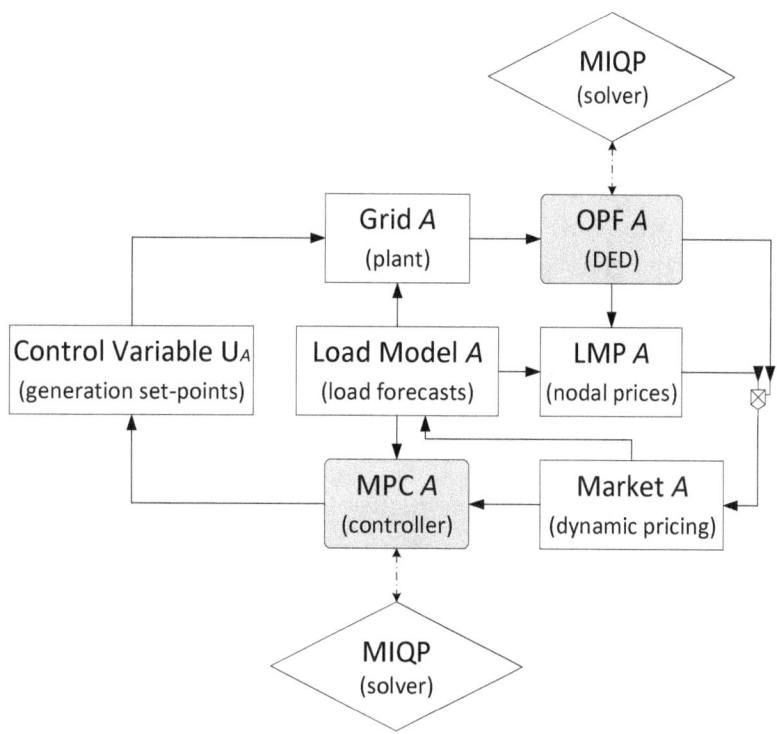

Figure 25.: A local feedback control loop for an exemplary grid unit A

As shown in the figure, a feedback control loop is formalized for an exemplary grid unit A, which also represents a general form of the proposed control loop for each individual grid unit. We use this feedback control loop to stabilize the local optimal dispatch of each local grid unit with integration of a correspondent local market unit. Besides the market and grid models proposed in Chapter 3, the local feedback control loop consists of the following 8 components in total. The grid model is divided into two components according the two-layer architecture, i.e. the topology module and the power flow module. A module of **Load Model** generates load profiles and executes load forecasting with given lead times. Details about the load forecasting model are provided in Chapter 6. The market model is as well separated into two components according to the perspective of LMP-based nodal prices and local market prices. A **MPC** module is formulated as the control strategy for the optimal power dispatch problem.

- **Grid:** refers to the *topology module* that describes a certain grid topology with parameters of the grid components, such as bus nodes, branches, generators and loads.

- **OPF:** refers to the *power flow module* that executes an optimal power flow run for the steady-state conditions of the given grid unit in terms of minimizing generation cost and loss.

- **Load Model:** refers to the *load module* that generates load profiles and load forecasts of load nodes within the given grid unit.

- **LMP:** refers to the *LMP market module* that calculates the LMP-based nodal prices within the given grid unit, which are determined based on OPF as the marginal cost.

- **Market:** refers to the *local market module* that represents a dynamic pricing mechanism, which determines local market clearing prices based on the LMP-based nodal prices and the output of **Load Model**.

- **MPC:** refers to the *MPC module* that optimizes predictive control plans for the control horizon $T \mapsto (k, \ldots, k + T - 1)$ in terms of optimal generation dispatch schedules.

- **MIQP:** refers to the module of an optimization solver. For our concrete use cases, a MIQP (mixed-integer quadratic programming) solver is applied.

- **Control Variable:** refers to generation set-points from the predictive control plans of the *MPC module* for the dispatch time step k.

4.4 OPTIMIZATION PROBLEMS FORMULATION

Within the above proposed control loop, there are mainly two optimization problems that need to be solved, i.e. a **market price optimization** and a **power dis-**

patch optimization. In this section, these two optimization problems are formulated based on the notation of the market and grid models in Chapter 3. The problem formulation for the market price optimization consists of an OPF formulation, a LMP formulation and a price stabilization formulation. The problem formulation for the power dispatch optimization refers mainly to a MPC formulation. These problem formulations are detailed in the following sub-sections.

4.4.1 *Market Price Optimization*

The OPF Formulation

As mentioned before, the basis of the market price optimization is the proposed market model in Chapter 3, which refers to a real-time price settlement model for LMP-based nodal prices. In order to determine the nodal prices, we need to solve the optimization problem expressed in (7) subject to the operational constraints of the correspondent grid unit. As a proper solution to this optimization problem, we show first that the OPF problem of the correspondent grid unit is related to this optimization problem through Lagrange duality; then, we solve the OPF problem instead to calculate the nodal prices. For this reason, we describe first the OPF formulation that is based on the general form of each grid unit in (10), but with more concrete cost functions in terms of the utility and cost expressions from the market modeling in Chapter 3. The objective of the following OPF formulation is to minimize the total costs for stabilizing the dynamic supply-demand balance within the grid unit. The optimization solution is a combination of generation and load changes within generation capacities, load shedding limits and transmission constraints to minimize total generation costs and maximize consumption utility values. As we consider only generation and load changes as the control variable, the prosumer part will be merged into producers and consumers as proposed in the grid modeling in Chapter 3. Then, we formulate the local OPF problem for a grid unit with n bus nodes as follows. For simplicity, the disutility function $c_j(\bullet)$ for the consumer at node j is considered as a negative representation of the utility function $u_j(\bullet)$; And the cost functions $c_i^P(\bullet)$ and $c_i^R(\bullet)$ for the consumed and reserve power generation of the producer at node i are assumed to be the same cost function $c_i(\bullet)$.

$$
\begin{aligned}
\min_{s,d} \quad \varphi(s-s_0, d-d_0) &\triangleq \sum_{i\in\mathbb{S}\cup\mathbb{P}} \left(c_i^P(s_i^P - s_{i0}^P) + c_i^R(s_i^R - s_{i0}^R) \right) \\
&\quad - \sum_{j\in\mathbb{D}\cup\mathbb{P}} \left(u_j(d_j - d_{j0}) - c_j(\Delta d_j - \Delta d_{j0}) \right) \\
&\triangleq \sum_{i\in\mathbb{S}\cup\mathbb{P}} c_i(s_i - s_{i0}) - \sum_{j\in\mathbb{D}\cup\mathbb{P}} u_j(d_j - d_{j0}) \quad (22)
\end{aligned}
$$

subject to

$$I = A_{nn} \cdot V \qquad (23)$$

$$\sum_{i \in \mathbb{S} \bigcup \mathbb{P}} (s_i - s_{i0}) = \sum_{j \in \mathbb{D} \bigcup \mathbb{P}} (d_j - d_{j0}) \qquad (24)$$

$$s_i = P_{i,g} + Q_{i,g}\imath, \ d_j = P_{j,l} + Q_{j,l}\imath, \ i,j \in [1,n], \ \imath = \sqrt{-1} \qquad (25)$$

$$s_j - d_j = P_{j,in} + Q_{j,in}\imath = V_j \cdot conj(I_j), \ j \in [1,n], \ \imath = \sqrt{-1} \qquad (26)$$

$$Re(\frac{d_j}{d_{j0}}) = Im(\frac{d_j}{d_{j0}}), \ j \in [1,n] \qquad (27)$$

$$s_i^{min} \leq s_i \leq s_i^{max}, \ 0 \leq s_i - s_{i0} \leq s_i^{max}, \ i \in \mathbb{S} \bigcup \mathbb{P} \qquad (28)$$

$$0 \leq d_j \leq d_{j0}, \ j \in \mathbb{D} \bigcup \mathbb{P} \qquad (29)$$

$$|V|^{min} \leq |V| \leq |V|^{max} \qquad (30)$$

$$|I_{ij}| = |a_{ij} \cdot (V_i - V_j)| \leq |I_{ij}|^{max}, \ i,j \in [1,n], i \neq j \qquad (31)$$

$$|\phi_{ij} + \alpha_{ij} \cdot ((s_i - s_{i0}) + (d_i - d_{i0}))| \leq f \cdot |\phi_{ij}|^{max}, \ i,j \in [1,n], i \neq j \qquad (32)$$

In the above OPF formulation, $\varphi(\bullet)$ in (22) stands for the OPF objective function. $s = [s_1, \ldots, s_i, \ldots, s_n]^{tr}$ and $d = [d_1, \ldots, d_j, \ldots, d_n]^{tr}$ are the supply and demand vector representing complex power generation and consumption. There is $s_i = 0$ or $d_j = 0$ if the bus node i has no producer or the bus node j has no consumer. Moreover, s_0 and d_0 denote the initial condition of supply and demand at each optimization step. Equality constraint (23) defines the nodal voltage-current relationship in the grid unit. Equality constraint (24) balances the power system with load and generation changes in equal quantities. Equations in (25) describe the definition of nodal complex power generation and consumption at each bus node. Equality constraint (26) expresses the conservation of electricity at each node, where $P_{j,in} + Q_{j,in}\imath$ defines the complex net power injection at node j. Equality constraint (27) forces the system to shed real and reactive load in equal proportions. Inequality constraints (28) and (29) describe the extent to which loads and generators can be controlled: nodal generations are limited within the capacity lower bound s_i^{min} and upper bound s_i^{max} and generation changes have to be set within the lower limit of a complete shut down and the upper limit of the maximum generation capacity s_i^{max}; In the case of load shedding, nodal loads have to be set between a complete shedding as the lower limit and the previous load d_{j0} as the upper limit. The final Inequality constraints (30), (31) and (32) define the measures used to identify violations for bus voltage, branch current and branch power flow, where a_{ij} is the admittance between node i and j, ϕ_{ij} is the transmission power flow from node j to node i, α_{ij} and f are denoted as a sensitivity factor and a percentage parameter for strengthening the transmission limitation. With the notation of the transmission power flow ϕ_{ij} and the notation of the grid model in Section 3.3.2, the nodal net power injection in (26) can be then defined by $P_{j,in} + Q_{j,in}\iota \triangleq \sum_{l\in\Omega_j} \phi_{jl}$.

The LMP Formulation

We assume that in a liberalized, market-oriented power system, different units are acting efficiently to maximize their own profits rather than only reduce the costs. Thus, consumers and producers will adjust their power usage and production level based on the real-time price information, respectively, i.e. according to the price-to-power mapping function $s_i(\lambda)$ (6) and $d_j(\lambda)$ (5) introduced in Chapter 3. Therefore, the operational goal in a liberalized power system is to determine the nodal price λ_i for each node i in the grid, in order to maximize the total benefit of the power system under all operational constraints. Using the welfare function based optimization model in (7) that is formulated for a real-time nodal price settlement in the market modeling section of Chapter 3, the LMP formulation for calculating the nodal prices is followed within the above described OPF framework. Regarding the above simplification of the disutility function $c_j(\bullet)$ and both cost functions $c_i^P(\bullet)$ and $c_i^R(\bullet)$, we first simplify accordingly the welfare optimization model (7) as follows. This simplification will not change the principle of the optimization model, but the calculation complexity for the later evaluation.

$$
\begin{aligned}
\max_\lambda W_{tot}(d(\lambda(t)), s(\lambda(t))) &\triangleq \sum_{j\in\mathbb{D}\cup\mathbb{P}} \left(u_j(d_j(\lambda_j(t))) - c_j(d_j(\lambda_j(t))) \right) \\
&\quad - \sum_{i\in\mathbb{S}\cup\mathbb{P}} \left(c_i^P(s_i(\lambda_i(t))) + c_i^R(s_i(\lambda_i(t))) \right) \\
&\triangleq \sum_{j\in\mathbb{D}\cup\mathbb{P}} u_j(d_j(\lambda_j(t))) - \sum_{i\in\mathbb{S}\cup\mathbb{P}} c_i(s_i(\lambda_i(t))) \quad (33)
\end{aligned}
$$

Then, we reformulate the above welfare maximization problem in (33) into a minimization problem as the LMP formulation that determines nodal prices, in order to determine the Lagrange duality between the OPF formulation and the LMP formulation.

$$
\min_\lambda \Phi(s(\lambda(t)), d(\lambda(t))) = \max_\lambda (-1) \cdot W_{tot}(d(\lambda(t)), s(\lambda(t))) \quad (34)
$$

$$
= \max_\lambda - \left(\sum_{j\in\mathbb{D}\cup\mathbb{P}} u_j(d_j(\lambda_j(t))) - \sum_{i\in\mathbb{S}\cup\mathbb{P}} c_i(s_i(\lambda_i(t))) \right) \quad (35)
$$

$$
= \min_\lambda \sum_{i\in\mathbb{S}\cup\mathbb{P}} c_i(s_i(\lambda_i(t))) - \sum_{j\in\mathbb{D}\cup\mathbb{P}} u_j(d_j(\lambda_j(t))) \quad (36)
$$

subject to the same constraints of the above OPF (23) – (32). The decision variable is then a vector of nodal prices $\lambda = [\lambda_1, \ldots, \lambda_n]^{tr}$.

In general, the classical OPF problem is non-convex [10]. However, Lavaei and Low [10] showed that the non-convex OPF problem can be either convexified by a certain convex relaxation technique or reformulated into its dual equivalence as a SDP-based (semidefinite programming) convex optimization. Based on that, convexity assumptions are made for both $c_i(\bullet)$ and $u_j(\bullet)$ in Chapter 3, in order to guarantee a convex representation of the above OPF problem formulation. The equality and inequality constraints regarding s and d, i.e. (26), (28) and (29), represent the power conservation rule and the generation and load limitations, which are a fulfilled condition by default. Therefore, we assume that there exists at least a feasible set of s and d, which meets these constraints. Thus, Slater's condition is satisfied for the convex OPF problem, which means that strong duality of the OPF problem holds. In other words, a strong Lagrange duality holds between the OPF formulation and the LMP formulation, which means that the Lagrange multiplier associated with the nodal balance condition (26) in the Lagrange dual problem of OPF, can express nodal prices for the correspondent LMP problem. Therefore, the following partial dual problem formulation to the above OPF problem provides an equivalent solution concept for the LMP problem.

$$\min_{\lambda} \ \Phi(s(\lambda), d(\lambda)) = \min_{s,d,\lambda} \ \varphi(s - s_0, d - d_0) - \sum_{j=1}^{n} \lambda_j \cdot (s_j - d_j - V_j \cdot conj(I_j)) \quad (37)$$

The Price Stabilization

As described in the market modeling in Chapter 3, if the market prices refer to the above determined LMP-based nodal prices, the proposed market-grid feedback system could be unstable in terms of a high price volatility [16]. For this reason, we presented a local market concept in Chapter 3, which is integrated in the price settlement model to stabilize the determined nodal prices λ. The local market concept refers to a calculation of stabilized local market prices π according to the determined nodal prices λ. The procedure of this price stabilization that has been described in Chapter 3 is outlined here again as follows:

1. Discretize the time t with discrete time step k that corresponds to the time interval $[k, k+1]$, $k \in \mathbb{N}_0$;

2. At time step k (the time interval $[k, k+1]$), the information of nodal prices $\{\lambda(k-1), \ldots, \lambda(k-1-T)\}$, local market prices $\{\pi(k-1), \ldots, \pi(k-1-T)\}$ and nodal demands $\{d(k-1), \ldots, d(k-1-T)\}$ are given;

3. Forecast nodal demands of the current time interval based on the past data: $\hat{d}_j(k) = LF_k(d_j(k-1), \ldots, d_j(k-1-T))$, $\forall j \in [1, n]$, where LF_k is a load forecasting function;

4. Calculate nodal prices of the current time interval in (7) with predicted demands: $\lambda(k) = \mathrm{argmax}_\lambda W_{tot}(\lambda|\hat{d}(k))$;

5. Calculate local market prices: $\pi(k) = \Pi_k(\lambda(k), \ldots, \lambda(k-1-T), \pi(k-1), \ldots, \pi(k-1-T))$, where Π_k is a stabilizing function that describes dynamic relationships between nodal prices and local market prices;

For simplicity, we consider only a time-invariant stabilizing function, i.e. $\Pi_k = \Pi$. As concrete functions, we employ two possible dynamic pricing principles proposed by Roozbehani et al. [15] to calculate the local market prices π, i.e. one is a *λ-based method* and the other one is a *subgradient-based method*. The application of these two methods for stabilizing π is illustrated as follows.

Let $\Pi^\lambda(\bullet)$ and $\Pi^{\mathcal{G}}(\bullet)$ denote the price stabilizing function of the λ-based method and the subgradient-based method, respectively. Then, the price dynamics of π are defined in both cases as follows:

λ-based method

$$\pi(k+1) = \Pi^\lambda(\lambda(k+1), \pi(k)) = \pi(k) + \gamma \cdot (\lambda(k+1) - \pi(k)) \tag{38}$$

Subgradient-based method

$$\pi(k+1) = \Pi^{\mathcal{G}}(\pi(k)) = \pi(k) + \gamma \cdot \mathcal{G}(\pi(k)) \tag{39}$$

where $\gamma > 0$ represents a weighting factor in case of the λ-based method or a step size in case of the subgradient-based method. $\mathcal{G}(\pi(k))$ is a subgradient direction regarding λ in the above optimization formulation (37), which refers to $-(s - d - V \cdot conj(I))$ as a vector representation. With the previous simplification of the disutility function $c_j(\bullet)$ and both cost functions $c_i^P(\bullet)$ and $c_i^R(\bullet)$, both price-to-power mapping functions (5) and (6) are reformulated in a vector representation by

$$d(\lambda(k)) = \underset{x(k)}{\mathrm{argmax}} \ u(x(k)) - \lambda(k) \cdot x(k) \tag{40}$$

$$s(\lambda(k)) = \underset{x(k)}{\mathrm{argmax}} \ \lambda(k) \cdot x(k) - c(x(k)) \tag{41}$$

Since $u(x(k))$ and $c(x(k))$ are both continuously differentiable, by solving both argmax, $d(\lambda(k))$ and $s(\lambda(k))$ can be expressed as the inverse functions of the first-order derivative of $u(x(k))$ and $c(x(k))$, i.e. there are $d(\lambda(k)) = \dot{u}^{-1}(\lambda(k))$ and

$s(\lambda(k)) = \dot{c}^{-1}(\lambda(k))$. Then, we put both inverse functions with price input $\pi(k)$ into $-(s - d - V \cdot conj(I))$, so that the subgradient direction is expressed by

$$\begin{aligned}
\mathcal{G}(\pi(k)) &= -s(\pi(k)) + d(\pi(k)) + V \cdot conj(I) \\
&= -\dot{c}^{-1}(\pi(k)) + \dot{u}^{-1}(\pi(k)) + V \cdot conj(I).
\end{aligned} \tag{42}$$

With these both pricing functions in (38) and (39), we can achieve the convergence of the price variable $\pi(k)$.

4.4.2 MPC Problem Formulation

As shown in Figure 25, after solving each iteration of the above market price optimization, the local market prices $\pi(k)$ — the output of the **Market** module — are used with a twofold challenge, i.e. 1) to determine nodal load adjustments for an update of the load forecasting model in the **Load Model** module; 2) to determine generation set-points with the problem formulation for a power dispatch optimization. For the load adjustment, we take the mapping function (5) into account, so that an adjustment on each nodal load d_j at time step k is calculated based on the mapping function (5): $d_j(k) = d_j(\pi_j(k)) = W_j^{-1}(\pi_j(k)) = \text{argmax}_{x_j(k), \Delta x_j(k)} \ u_j(x_j(k)) - c_j(\Delta x_j(k)) - \lambda_j(t) \cdot x_j(k), \forall j \in [1, n]$. However, the updated $d_j(k)$ will be first applied in the next control iteration at time $k+1$.

In addition to the load adjustment, we need to determine matched generation set-points, in order to force the power imbalance to zero. Achieving a balance between the adjusted load and the desired generation, is a control problem. In this sub-section, we focuses on the problem formulation of a feedback controller that takes nodal dynamic prices of the local market as input, and determines set-point adjustments of generators as output (see Figure 25), in order to hold the supply-demand balance. This feedback control problem is formulated as a MPC-based generation and load dispatch problem, which maximizes the total social welfare for determining generation set-point adjustments and load shedding adjustments as control variables, in the meantime satisfies all the required constraints.

Before we proceed with the MPC problem formulation as a controller synthesis, we specify first the state-space representation of the local grid model considering both intra- and inter-transmission power flow. For simplicity, we assume that nodal dynamics of a single grid unit G^i with n_i bus nodes can be expressed as a discrete time-invariant system of the grid's active powers in a vector representation as follows:

$$\begin{bmatrix} \bar{s}_i(k+1) \\ \bar{d}_i(k+1) \end{bmatrix} = \begin{bmatrix} \bar{s}_i(k) \\ \bar{d}_i(k) \end{bmatrix} + \begin{bmatrix} \Delta\bar{s}_i(k) \\ \Delta\bar{d}_i(k) \end{bmatrix} \tag{43}$$

$$\bar{P}_{i,in}(k) = \begin{bmatrix} \bar{I}_{n_i} & -\bar{I}_{n_i} \end{bmatrix} \cdot \begin{bmatrix} \bar{s}_i(k) \\ \bar{d}_i(k) \end{bmatrix} + \begin{bmatrix} \bar{I}_{n_i} & -\bar{I}_{n_i} \end{bmatrix} \cdot \begin{bmatrix} \Delta\bar{s}_i(k) \\ \Delta\bar{d}_i(k) \end{bmatrix} \tag{44}$$
$$+ \bar{w}_{i,in}(k) + \bar{w}_{i,out}(k)$$

where $\bar{x}_i = \begin{bmatrix} \bar{s}_i \\ \bar{d}_i \end{bmatrix} \in \mathbb{R}^{2n_i}$ is the state vector representing nodal generation conditions

$\bar{s}_i = [s_1,\ldots,s_{n_i}]^{tr}$ and load conditions $\bar{d}_i = [d_1,\ldots,d_{n_i}]^{tr}$; $\bar{u}_i = \begin{bmatrix} \Delta\bar{s}_i \\ \Delta\bar{d}_i \end{bmatrix} \in \mathbb{R}^{2n_i}$ is the control vector representing nodal generation adjustments $\Delta\bar{s}_i = [\Delta s_1,\ldots,\Delta s_{n_i}]^{tr}$ and load adjustments $\Delta\bar{d}_i = [\Delta d_1,\ldots,\Delta d_{n_i}]^{tr}$; $\bar{y}_i = \bar{P}_{i,in} \in \mathbb{R}^{n_i}$ is the output vector representing nodal power injections $\bar{P}_{i,in} = \begin{bmatrix} P^i_{1,in},\ldots,P^i_{n_i,in} \end{bmatrix}^{tr}$; $\bar{w}_{i,in} \in \mathbb{R}^{n_i}$ and $\bar{w}_{i,out} \in \mathbb{R}^{n_i}$ are the vector of interconnection input power flows $\bar{w}_{i,in} = \begin{bmatrix} P^i_{in,1},\ldots,P^i_{in,n_i} \end{bmatrix}^{tr}$ and output power flows $\bar{w}_{i,out} = \begin{bmatrix} P^i_{out,1},\ldots,P^i_{out,n_i} \end{bmatrix}^{tr}$, respectively; $P^i_{in,j}$ and $P^i_{out,j}$, $j \in [1,n_i]$, denote the active power flow into and out of the grid unit G^i through the node j.

As the output expression in (44) shown that the influence of nodal generation and load adjustments on nodal power injections $\bar{P}_{i,in}$ is reflected in both intra-transmission power flows and inter-transmission power flows. In this chapter, we consider only the system modeling of a local market-grid coupling for a single grid unit, the inter-transmission between different grid units are neglected, which means $\bar{w}_{i,in}(k) = \bar{0}$ and $\bar{w}_{i,out}(k) = \bar{0}$. The inter-transmission will be taken into account for the distributed market-grid coupling in the next chapter. Then, in the context of MPC for determining the control vector $\bar{u}_i$, we need to define a local grid objective function $\Phi_{i,local}(\tilde{\bar{x}}_i(k+1|s), \tilde{\bar{u}}_i(k|s), \tilde{\bar{y}}_i(k|s))$ for the MPC optimization problem, where s behind the vertical bar indicates the current optimization iteration, the tilde over the vector variables indicates variables over the prediction horizon T, e.g. $\tilde{\bar{u}}_i(k|s) \equiv \begin{bmatrix} (\bar{u}_i(k|s))^{tr},\ldots,(\bar{u}_i(k+T-1|s))^{tr} \end{bmatrix}^{tr}$. The MPC problem for a single grid unit G^i with the objective of maximizing the total social welfare is then formulated as:

$$\max_{\tilde{\bar{v}}_i(k|s)} \Phi_{i,local}(\tilde{\bar{x}}_i(k+1|s), \tilde{\bar{u}}_i(k|s), \tilde{\bar{y}}_i(k|s)) \tag{45}$$

where $\tilde{v}_i(k|s) \equiv \{\tilde{x}_i(k+1|s), \tilde{u}_i(k|s), \tilde{y}_i(k|s)\}$ represents the set of all optimization variables. The objective function is subject to the above state-space representation (43) – (44) and the following constraints:

$$\|\Delta \bar{s}_i(k)\|_1 = \|\Delta \bar{d}_i(k)\|_1 \tag{46}$$

$$\bar{r}_i^{down} \leq \Delta \bar{s}_i(k) \leq \bar{r}_i^{up} \tag{47}$$

$$\bar{s}_i^{min} \leq \bar{s}_i(k) \leq \bar{s}_i^{max} \tag{48}$$

$$\bar{P}_{i,in}^{min} \leq \bar{P}_{i,in}(k) \leq \bar{P}_{i,in}^{max} \tag{49}$$

Equality constraint (46) balances total generation adjustments with total load changes. Inequality constraint (47) denotes the down and up ramp rates of generators. Inequality constraints (48) and (49) indicate upper and lower bounds of the nodal generation and the intra-transmission power flow. In order to determine the control vector $\bar{u}_i(k|s)$ based on the above MPC formulation, we introduce the following social welfare expression into the MPC objective function (45).

$$
\begin{aligned}
&\Phi_{i,local}(\tilde{x}_i(k+1|s), \tilde{u}_i(k|s), \tilde{y}_i(k|s)) \\
&= \sum_{l=0}^{T-1} (\bar{\pi}_i(k+l))^{tr} \cdot \begin{bmatrix} \bar{x}_i(k+l+1|s) \\ \bar{u}_i(k+l|s) \\ \bar{y}_i(k+l|s) \end{bmatrix} \\
&\quad - \begin{bmatrix} \bar{x}_i(k+l+1|s) \\ \bar{u}_i(k+l|s) \\ \bar{y}_i(k+l|s) \end{bmatrix}^{tr} \cdot Q_i \cdot \begin{bmatrix} \bar{x}_i(k+l+1|s) \\ \bar{u}_i(k+l|s) \\ \bar{y}_i(k+l|s) \end{bmatrix} \\
&\quad - R_i^{tr} \cdot \begin{bmatrix} \bar{x}_i(k+l+1|s) \\ \bar{u}_i(k+l|s) \\ \bar{y}_i(k+l|s) \end{bmatrix} \tag{50}
\end{aligned}
$$

$$
\begin{aligned}
&= \sum_{l=0}^{T-1} \left((\bar{\pi}_i(k+l))^{tr} - R_i^{tr} \right) \cdot \begin{bmatrix} \bar{x}_i(k+l+1|s) \\ \bar{u}_i(k+l|s) \\ \bar{y}_i(k+l|s) \end{bmatrix} \\
&\quad - \begin{bmatrix} \bar{x}_i(k+l+1|s) \\ \bar{u}_i(k+l|s) \\ \bar{y}_i(k+l|s) \end{bmatrix}^{tr} \cdot Q_i \cdot \begin{bmatrix} \bar{x}_i(k+l+1|s) \\ \bar{u}_i(k+l|s) \\ \bar{y}_i(k+l|s) \end{bmatrix} \tag{51}
\end{aligned}
$$

where we assume a linear relationship between the total revenue and electricity power units in terms of generation, load and injection. The linearity is expressed as a price vector $\bar{\pi}_i(k) = [(\bar{\pi}_{i,g}(k))^{tr}, (\bar{\pi}_{i,l}(k))^{tr}, (\bar{\pi}_{i,g}(k))^{tr}, (\bar{\pi}_{i,l}(k))^{tr}, (\bar{\pi}_{i,in}(k))^{tr}]^{tr}$, in which $\bar{\pi}_{i,g}(k)$ denotes nodal local market prices for the generation power, $\bar{\pi}_{i,l}(k)$ denotes nodal utility multipliers (a simplification of the utility function as defined in the market modeling in Chapter 3) for the load power, $\bar{\pi}_{i,in}(k) = \bar{\pi}_{i,g}(k)$ denotes "toll charges" for the injected power as revenue of network players. Q_i and R_i are weighting matrices consisting of sub-matrices that are constructed as follows and correspond to the costs of state variables (generation and load conditions), control variables (generation and load adjustments) and output variables (nodal power injections). Both matrices have to be appropriately tuned to weight the costs of state, control and output variables for the objective of the total social welfare.

$$Q_i = \begin{bmatrix} Q_{xi} & & \\ & Q_{ui} & \\ & & Q_{yi} \end{bmatrix} \tag{52}$$

$$R_i = \begin{bmatrix} R_{xi} \\ R_{ui} \\ R_{yi} \end{bmatrix} \tag{53}$$

where Q_{xi} and R_{xi} correspond to the costs of the generation and load power, Q_{ui} and R_{ui} correspond to the costs of the generation and load adjustment, and Q_{yi} and R_{yi} correspond to the costs of the power injection. From the control point of view, the optimization objective is the minimization of costs of the generation and load adjustment for maximizing the total social welfare, so that the costs related to $\bar{x}_i$ and $\bar{y}_i$ will not be further considered, which means $Q_{xi} = [0]$, $R_{xi} = [0]$, $Q_{yi} = [0]$ and $R_{yi} = [0]$. Therefore, for the following evaluation, we focus only on the matrices Q_{ui} and R_{ui}.

4.5 TEST AND EVALUATION

In this section, we introduce evaluation scenarios for different grid models that are constructed based on IEEE bus test cases and present simulation-based numerical results to demonstrate the capability of the proposed MPC-based feedback control system depicted in Figure 25. As this feedback control system is designed for a local market-grid coupling, the evaluation in all use cases comprises two aspects, namely market price and power dispatch.

For a comparison between evaluation scenarios, we set for all grid test cases with the following parametrization and configuration. For each generator bus, we assume that its aggregate cost is estimated by a quadratic cost function $c_i(x) = \alpha_2^i \cdot x^2 + \alpha_1^i \cdot x + \alpha_0^i$. $\alpha_{0,1,2}^i$ are cost coefficients for polynomial cost functions from the IEEE test case and $\alpha_2^i > 0$ ensures the convexity. We further assume that for each load bus, the consumers' total utility value is expressed by a logarithmic utility function $u_j(x) = \beta^j \cdot \ln x$, so that the inverse function $d_j(\pi_j)$ of power load can be expressed as $d_j = \beta^j / \pi_j$, where π_j indicates nodal local market prices. The logarithmic function implies normal consumers' risk-averse behavior in microeconomics, which means that most of the consumers prefer a fixed budget for their energy bills [8, 19]. Furthermore, different values of the coefficient β^j represent different consumer/load types. As consumer/load at each load bus represents an aggregate load, we notice in our initial test that different load types don't affect the optimization results significantly. Therefore, for simplicity, we simulate all load buses with the same consumer/load type, i.e. $\forall j : \beta^j = 1500$. Besides, for the pricing function, we employ the λ-based method expressed in (38), which is also proposed by Roozbehani et al. [17] for stabilizing the local market price dynamics $\pi(k)$ based on the knowledge of LMPs $\lambda(k)$, i.e. $\pi(k+1) = \pi(k) + \gamma \cdot (\lambda(k+1) - \pi(k))$, where $\gamma = 0.05$.

Both optimizations in each local control loop are conducted by a MIQP (mixed-integer quadratic programming) solver in Matlab. To make a k-step simulation, we need the load/consumption information of those k steps. For more realistic nodal loads with stochastic aspects, we simulate nodal power loads $d_j(k)$ with an additive white noise $\epsilon_j \sim \mathcal{N}\left(0, \sigma_j^2\right)$ and cap them with lower and upper bounds (d_{min} and d_{max}): $d_j(k) = \max(d_{min}, \min(d_{max}, \beta^j / \pi_j(k) \cdot (1 + \epsilon_j)))$, where $d_{min} = 0$, d_{max} is assigned by the power demand from the correspondent IEEE test case, and $\sigma_j = 1/3$.

4.5.1 IEEE 14 Bus Test Case

The first use case refers to a small-size grid network, the IEEE 14 Bus test case[1], which represents a simple approximation of the American Electric Power system (in the Midwestern US) as of February 1962 in form of a 14-bus system and presents data of grid components and network elements in form of a bus table, a generator table and a branch table. This IEEE 14-bus system contains 14 buses, 5 generators, 20 lines and 11 loads. The following simulation demonstrates how the proposed feedback control system works on this IEEE 14-bus system.

The simulation are conducted with 200 time steps. The optimization results within the feedback control framework are depicted in the following 5 figures. In order to compare the prediction impact of the optimization results, the prediction horizon of the MPC has been set with 3 options, i.e. $T = 1$, $T = 5$ and $T = 10$. Figure 26, Figure 27 and Figure 28 show the nodal load adjustments Δd of the 14-bus system with prediction horizon $T = 1$, $T = 5$ and $T = 10$, respectively. We notice that the load adjustments are more damped in the case of $T = 10$ compared to $T = 1, 5$, which implies that the more future information the MPC perceives, the optimization objective, i.e. the load dispatch Δd, can be then stabilized with less oscillation. Figure 29 depicts the determined local market price π of the 14-

---

1 http://icseg.iti.illinois.edu/ieee-14-bus-system/

bus system in cases $T = 1$, $T = 5$ and $T = 10$. As shown in the figure, the price π in case $T = 1$ remains almost constant at 41 (monetary units) from the 75-th time step, which is stabilized with less oscillation in comparison to the case of $T = 1, 5$. However, in case $T = 1$ and $T = 5$, the stable price π is similar and amounts to ca. 40.5 beginning at the 110-th time step. The difference between the stable prices of all three cases are not so huge, which means that the impact of different prediction horizons on the market price stability for the IEEE 14-bus system is not so significant.

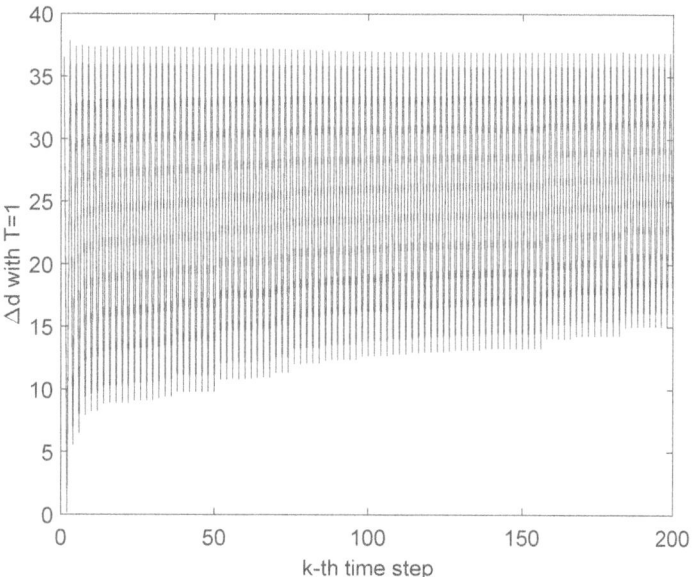

Figure 26.: Evolution of nodal load adjustments Δd with $T = 1$ on the IEEE 14-bus system

Now, we analyze the execution time of solving the optimization problems on the IEEE 14-bus system. Figure 30 shows the run time data of the 200 simulation time steps for all 3 prediction horizon cases, i.e. $T = 1$, $T = 5$ and $T = 10$. We notice that the average run time of each simulation time step is about $4\,s$ in case $T = 1$, $8\,s$ in case $T = 5$ and $10\,s$ in case $T = 10$. For a small-size grid network, such as the IEEE 14-bus system, the execution time of the proposed optimizations in each case is less than 1 minute, which answers our question left in Chapter 3 that the above MPC-based control problem for small-size grid networks can be solved within the same time scale as the set requirement of the proposed market-grid coupling.

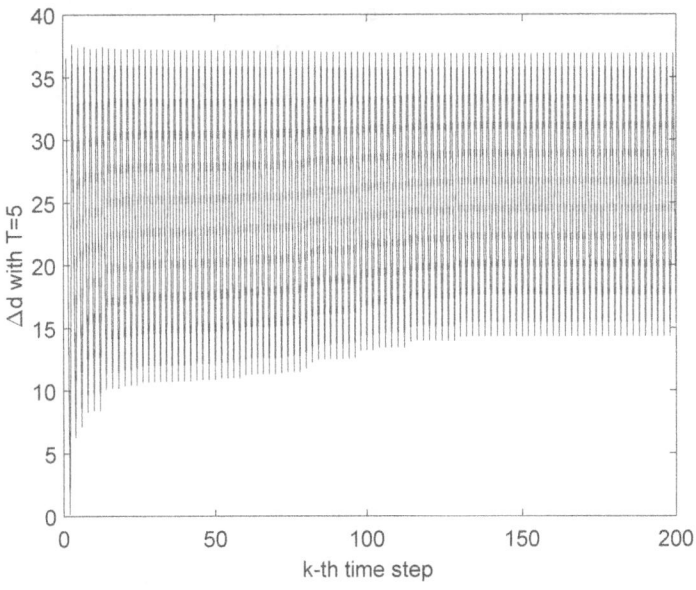

Figure 27.: Evolution of nodal load adjustments Δd with $T = 5$ on the IEEE 14-bus system

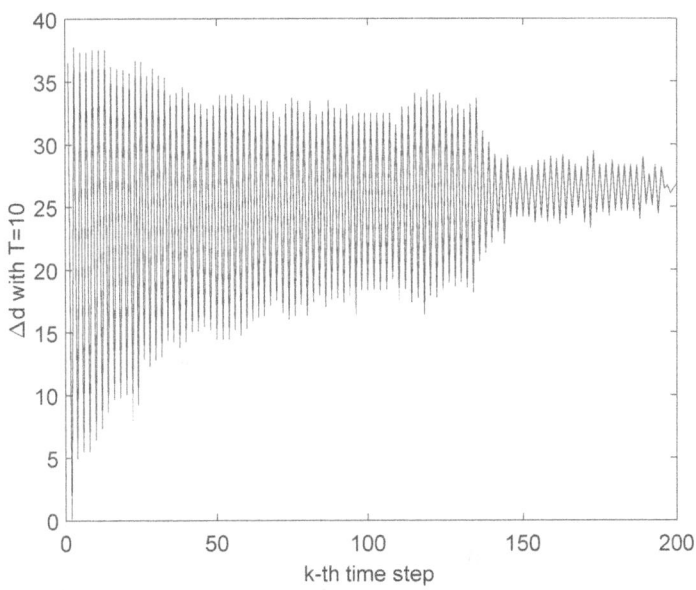

Figure 28.: Evolution of nodal load adjustments Δd with $T = 10$ on the IEEE 14-bus system

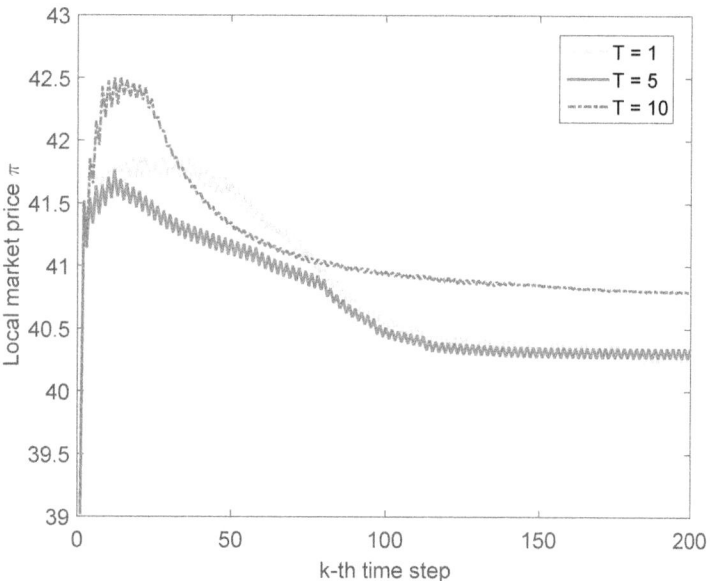

Figure 29.: Evolution of local market prices π with $T = 1, 5, 10$ on the IEEE 14-bus system

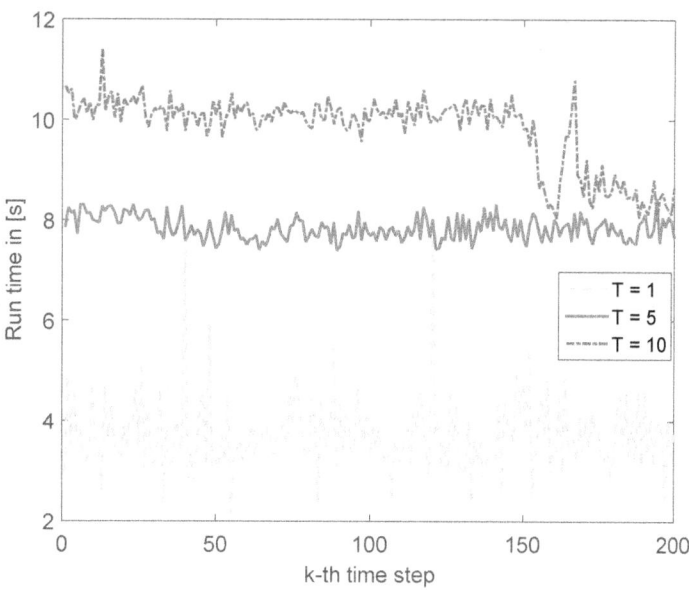

Figure 30.: Evolution of simulation run time with $T = 1, 5, 10$ on the IEEE 14-bus system

4.5.2 *IEEE 118 Bus Test Case*

The second use case refers to a medium-size grid network, the IEEE 118 Bus test case[2], which represents a simple approximation of the American Electric Power system (in the U.S. Midwest) as of December 1962 in form of a 118-bus system and presents data of grid components and network elements in form of a bus table, a generator table and a branch table. This IEEE 118-bus system contains 118 buses, 19 generators, 35 synchronous condensers, 177 lines, 9 transformers and 91 loads. The following simulation demonstrates an application of the proposed feedback control system to this IEEE 118-bus system.

The simulation results with 200 time steps are shown in the following 5 figures. The prediction horizon of the MPC has been set with 3 options, i.e. $T = 1$, $T = 5$ and $T = 10$. Figure 31, Figure 32 and Figure 33 show the nodal load adjustments Δd of the 118-bus system with prediction horizon $T = 1$, $T = 5$ and $T = 10$, respectively. We notice that the load adjustments are more damped in the case of $T = 10$ compared to $T = 1, 5$, which implies that the more future information the MPC perceives, the load dispatch Δd can be then stabilized with less oscillation. Figure 34 depicts the local market price π of the 118-bus system in cases $T = 1$, $T = 5$ and $T = 10$. As shown in the figure, the price π in case $T = 1$ remains almost constant at 40 (monetary units) that is much more stabilized than in case $T = 5, 10$. However, in case $T = 5$ and $T = 10$, the latest stable price π is about 37 and 23, which are much lower than in case $T = 1$, one possible reason for this phenomenon is that the price π in case $T = 5, 10$ is determined by more optimized load dispatch Δd for the local consumer.

Now, we analyze the execution time of solving the optimization problems in the control loop of the IEEE 118-bus system. Figure 35 shows the run time data of the 200 simulation time steps for all 3 prediction horizon cases, i.e. $T = 1$, $T = 5$ and $T = 10$. We notice that the average run time of each simulation time step is about 25 s in case $T = 1$, while the average run time in both cases $T = 5$ and $T = 10$ amounts to about 32 s. In each case, the execution time is less than 1 minute, which again answers our question left in Chapter 3 that the MPC-based control problem for a medium-size grid network, such as the IEEE 118-Bus system, can be solved within the same time scale as the set requirement of the proposed market-grid coupling.

---

2 http://icseg.iti.illinois.edu/ieee-118-bus-system/

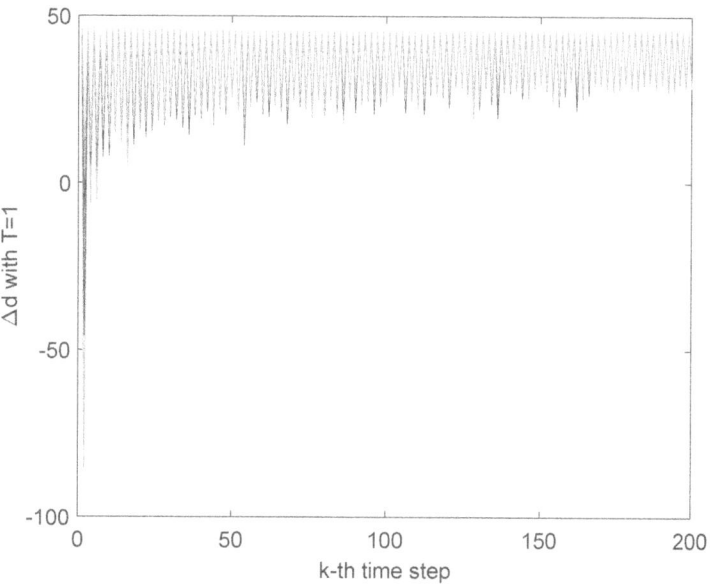

Figure 31.: Evolution of nodal load adjustments Δd with $T = 1$ on the IEEE 118-bus system

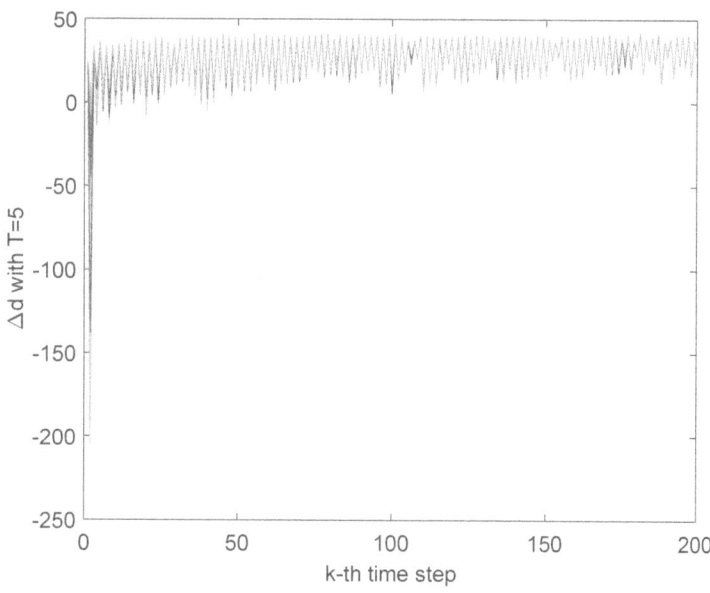

Figure 32.: Evolution of nodal load adjustments Δd with $T = 5$ on the IEEE 118-bus system

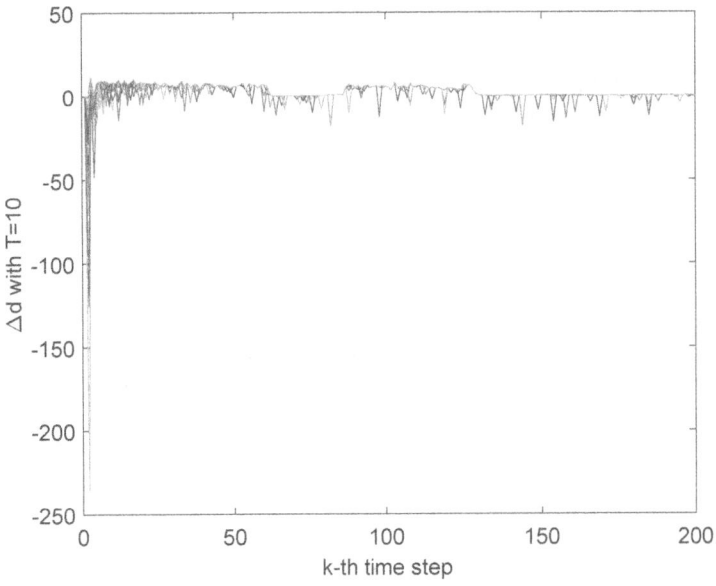

Figure 33.: Evolution of nodal load adjustments Δd with $T = 10$ on the IEEE 118-bus system

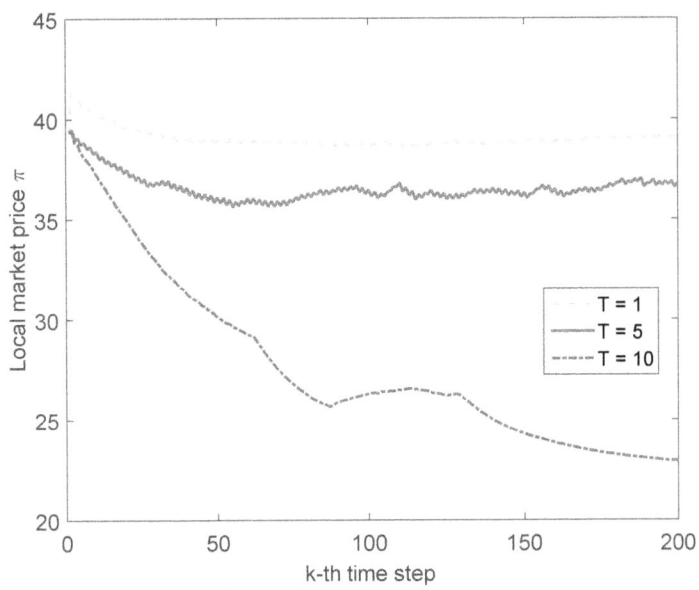

Figure 34.: Evolution of local market prices π with $T = 1, 5, 10$ on the IEEE 118-bus system

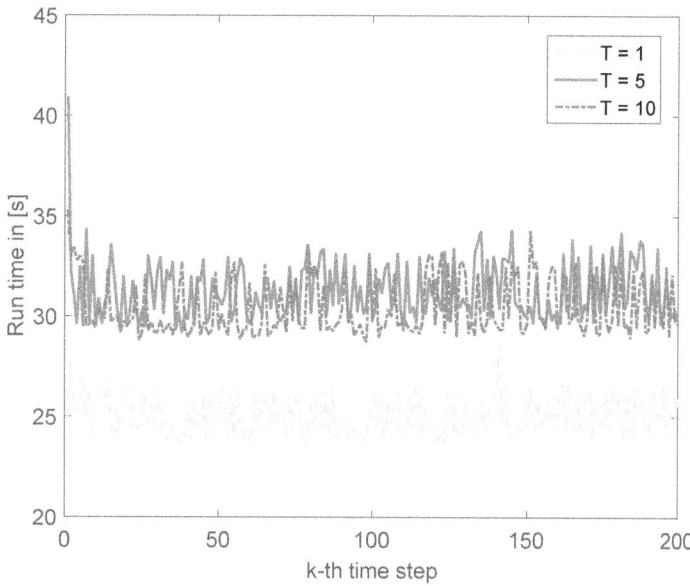

Figure 35.: Evolution of simulation run time with $T = 1, 5, 10$ on the IEEE 118-bus system

4.5.3 IEEE 300 Bus Test Case

As the third use case, we utilize the IEEE 300 Bus test case[3] to represent a large-size grid network. This test case was developed by the IEEE Test Systems Task Force led by Mike Adibi in 1993 and presents data of grid components and network elements in form of a bus table, a generator table and a branch table. This IEEE 300-bus system contains 300 buses, 69 generators, 410 lines and 195 loads. We notice a sub-grid within this IEEE 300-bus system, which represents an isolated microgrid disconnected to the other bus nodes. Due to this disconnected grid part, an OPF on the entire 300-bus system is always violated. For this reason, we conduct the simulation of the proposed feedback control system on the modified IEEE 300-bus system that eliminates the correspondent buses and branches of that disconnected sub-grid. Then, the following simulation demonstrates an application of the proposed feedback control system to this modified IEEE 300-bus system that, in fact, contains 265 interconnected buses.

The simulation results with 200 time steps are shown in the following 5 figures. The same as the other use cases, the prediction horizon of the MPC has been set with 3 options, i.e. $T = 1$, $T = 5$ and $T = 10$. Figure 36, Figure 37 and

---

3 http://icseg.iti.illinois.edu/ieee-300-bus-system/

Figure 38 show the nodal load adjustments Δd of the modified 300-bus system with prediction horizon $T = 1$, $T = 5$ and $T = 10$, respectively. We notice that the load adjustments in all 3 prediction horizon cases are damped well with just small load dispatch actions around the zero except for the beginning phase that refers to a MPC initialization. Figure 39 depicts the local market price π of the modified 300-bus system in cases $T = 1$, $T = 5$ and $T = 10$. As shown in the figure, the price π in case $T = 1$ remains almost constant at 30 (monetary units) that is much faster convergent than in case $T = 5, 10$. However, in case $T = 5$ and $T = 10$, the price π within the first 200 time steps is still in a stabilization phase, where the price π converges slowly to a stable price, i.e. about 20 in case $T = 5$ and 15 in case $T = 10$. That means that the stable price in case $T = 5, 10$ is much lower than in case $T = 1$, one possible reason for this phenomenon is similar to the simulation test for the above 118-bus system that the price π in case $T = 5, 10$ is determined by more optimized load dispatch Δd for the local consumer. In Chapter 5, we will utilize the IEEE 300-bus system again and disaggregate the entire 300-bus grid network into 4 grid units, for which we will show how the interconnected grid units can benefit from their own local market for a price advantage.

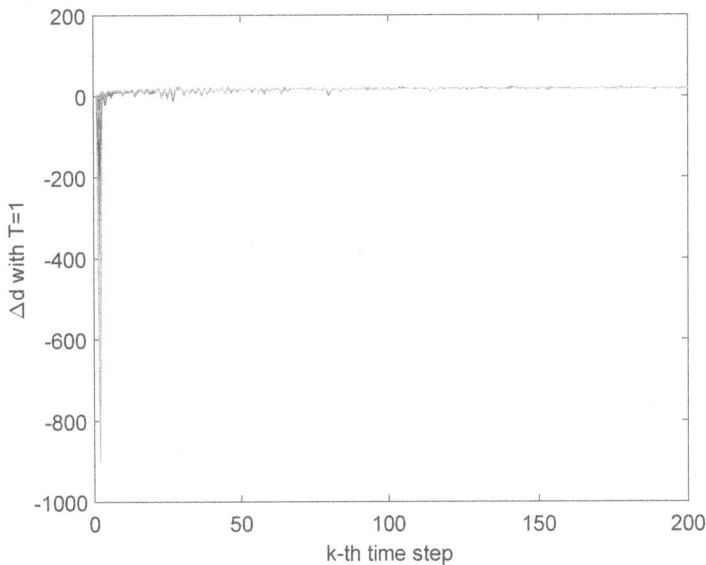

Figure 36.: Evolution of nodal load adjustments Δd with $T = 1$ on the modified IEEE 300-bus system

Moreover, similar to the previous use cases, we analyze the execution time of solving the optimization problems in the control loop of the modified IEEE 300-bus system. Figure 40 shows the run time data of the 200 simulation time steps for all 3 prediction horizon cases, i.e. $T = 1$, $T = 5$ and $T = 10$. We notice that

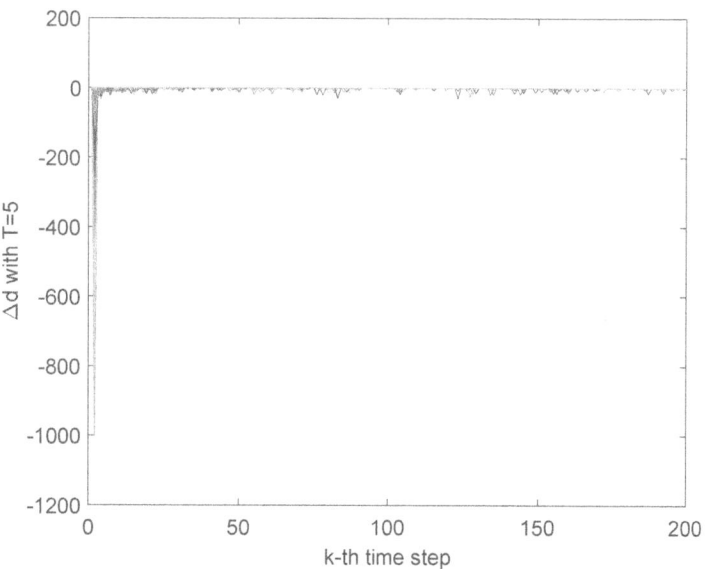

Figure 37.: Evolution of nodal load adjustments Δd with $T = 5$ on the modified IEEE 300-bus system

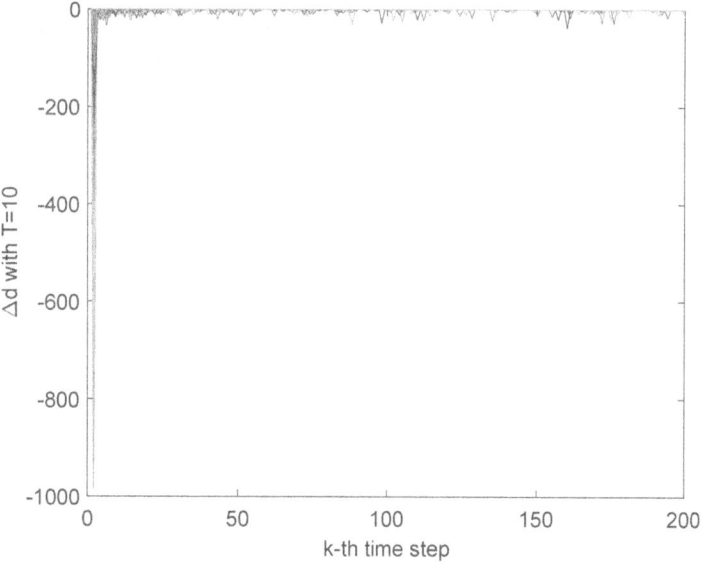

Figure 38.: Evolution of nodal load adjustments Δd with $T = 10$ on the modified IEEE 300-bus system

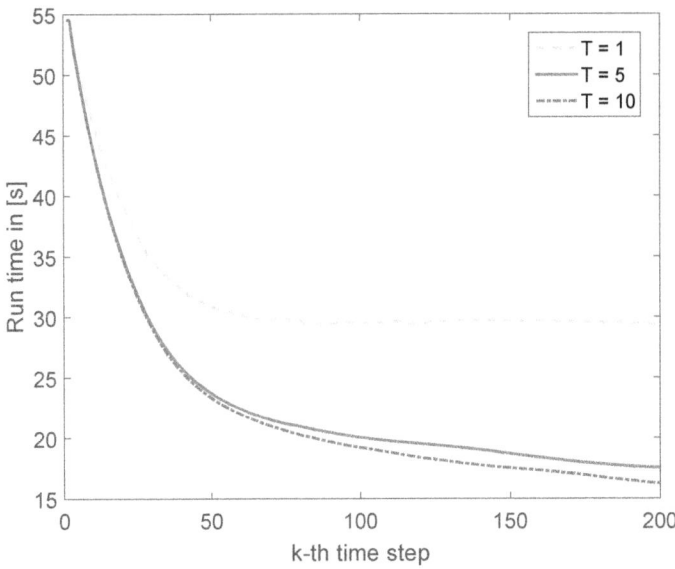

Figure 39.: Evolution of local market prices π with $T = 1, 5, 10$ on the modified IEEE 300-bus system

the average run time of each simulation time step is about $50\ s$ in case $T = 1$, while the average run time in both cases $T = 5$ and $T = 10$ amounts to about $95\ s$ or $100\ s$. For all 3 prediction horizon cases, the average execution time is less than $110\ s$, which again meets the time response requirement of the proposed market-grid coupling that the MPC-based control problem also for large-size grid networks can be solved within a time scale of minutes.

4.6 CONCLUSION

In this chapter, we further investigated and analyzed the previously-defined market-grid coupling in two different directions, i.e. a simulation framework and a control system. First, we proposed a co-simulation framework that realizes a market-grid coupling for studying the load influence on the market price in consideration of physical grid dynamics and constraints. In this co-simulation framework, the proposed market-grid coupling has been only analyzed with the aspect of the grid impact on the market price. We, then, presented the system modeling of a MPC-based closed-loop feedback control system for an interoperable control between the market and the grid, in which a market price optimization and a power dispatch optimization are performed concurrently. In addition, we provided necessary propositions and correspondent proofs for the formulation of proposed

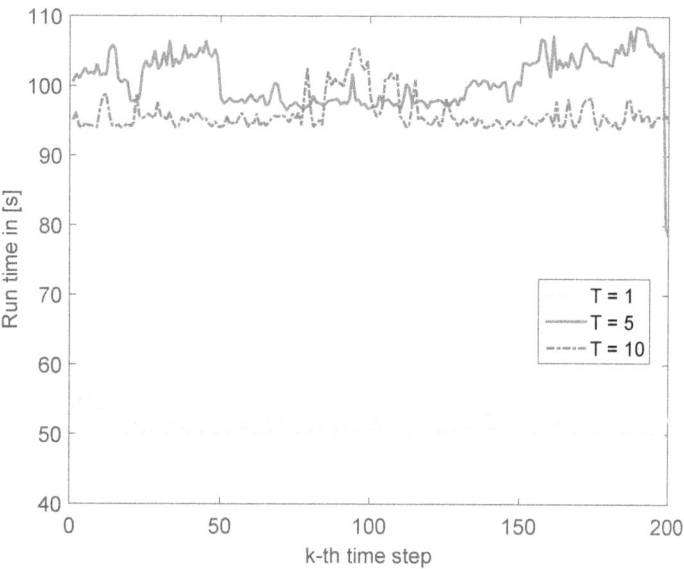

Figure 40.: Evolution of simulation run time with $T = 1, 5, 10$ on the modified IEEE 300-bus system

optimization problems. The problem formulation of the control system has been focused on a coupling model with a single grid unit and its correspondent local market. Finally, we introduced 3 IEEE bus system test cases as evaluation use cases for analyzing the proposed MPC-based feedback control system. For each use case, we demonstrated the obtained simulation-based numerical results with focus on the market price, the power load dispatch and the optimization execution time. The simulation results showed that this MPC-based control system for a local market-grid coupling provides a market price stability and a power dispatch stability to varying degrees for different grid models. The run time evaluation confirmed that in each case the optimization based on the proposed control system meets the time response requirement (within a time scale of minutes) of the proposed market-grid coupling. In the next chapter, we introduce a distributed control architecture for extending the centralized MPC problem of a local market-grid coupling to a distributed MPC problem of a distributed market-grid coupling model.

Bibliography

[1] D. Aliprantis, S. Penick, L. Tesfatsion, and Huan Zhao. Integrated retail and wholesale power system operation with smart-grid functionality. In *Power and Energy Society General Meeting, 2010 IEEE*, pages 1–8, July 2010. doi: 10. 1109/PES.2010.5589594.

[2] K. Anderson, J. Du, A. Narayan, and A. El Gamal. Gridspice: A distributed simulation platform for the smart grid. *Industrial Informatics, IEEE Transactions on*, 10(4):2354–2363, Nov 2014. doi: 10.1109/TII.2014.2332115.

[3] David Angeli. Economic model predictive control. *Encyclopedia of Systems and Control*, pages 337–343, 2015.

[4] D.P. Chassin, K. Schneider, and C. Gerkensmeyer. Gridlab-d: An open-source power systems modeling and simulation environment, 2008.

[5] Yong Ding, M.A. Neumann, M. Budde, M. Beigl, P.G. Da Silva, and Lin Zhang. A control loop approach for integrating the future decentralized power markets and grids. In *Smart Grid Communications (SmartGridComm), 2013 IEEE International Conference on*, pages 588–593, Oct 2013. doi: 10.1109/ SmartGridComm.2013.6688022.

[6] Yong Ding, Axel Morawietz, and Michael Beigl. Investigation of a grid-driven real-time pricing in a simulation environment. In *in Proceedings of IEEE International Energy Conference (ENERGYCON 2016)*, Leuven, Belgium, April 2016. IEEE.

[7] Yong Ding, Erwin Stamm, and Michael Beigl. Using model predictive control to interact distributed power markets and grids. In *in Proceedings of IEEE International Energy Conference (ENERGYCON 2016)*, Leuven, Belgium, April 2016. IEEE.

[8] Mehdi Farsi. Risk aversion and willingness to pay for energy efficient systems in rental apartments. *Energy Policy*, 38(6):3078–3088, 2010.

[9] Carlos E. Garcia, David M. Prett, and Manfred Morari. Model predictive control: Theory and practice - a survey. *Automatica*, 25(3):335 – 348, 1989. ISSN 0005-1098. doi: http://dx.doi.org/10.1016/0005-1098(89) 90002-2. URL http://www.sciencedirect.com/science/article/pii/0005109889900022.

[10] J. Lavaei and S.H. Low. Zero duality gap in optimal power flow problem. *Power Systems, IEEE Transactions on*, 27(1):92–107, Feb 2012. ISSN 0885-8950. doi: 10.1109/TPWRS.2011.2160974.

[11] Hongyan Li and L. Tesfatsion. The ames wholesale power market test bed: A computational laboratory for research, teaching, and training. In *Power Energy Society General Meeting, 2009. PES '09. IEEE*, pages 1–8, July 2009. doi: 10.1109/PES.2009.5275969.

[12] F. Milano, L. Vanfretti, and J.C. Morataya. An open source power system virtual laboratory: The psat case and experience. *Education, IEEE Transactions on*, 51(1):17–23, Feb 2008. doi: 10.1109/TE.2007.893354.

[13] James Blake Rawlings and David Q Mayne. *Model predictive control: Theory and design*. Nob Hill Pub., 2009.

[14] J.B. Rawlings, D. Angeli, and C.N. Bates. Fundamentals of economic model predictive control. In *Decision and Control (CDC), 2012 IEEE 51st Annual Conference on*, pages 3851–3861, Dec 2012. doi: 10.1109/CDC.2012.6425822.

[15] M. Roozbehani, Munther Dahleh, and S. Mitter. Dynamic pricing and stabilization of supply and demand in modern electric power grids. In *Smart Grid Communications (SmartGridComm), 2010 First IEEE International Conference on*, pages 543–548, Oct 2010. doi: 10.1109/SMARTGRID.2010.5621994.

[16] M. Roozbehani, Munther Dahleh, and S. Mitter. On the stability of wholesale electricity markets under real-time pricing. In *Decision and Control (CDC), 2010 49th IEEE Conference on*, pages 1911–1918, Dec 2010. doi: 10.1109/CDC.2010.5718173.

[17] M. Roozbehani, M.A. Dahleh, and S.K. Mitter. Volatility of power grids under real-time pricing. *Power Systems, IEEE Transactions on*, 27(4):1926–1940, Nov 2012. doi: 10.1109/TPWRS.2012.2195037.

[18] Junjie Sun, Leigh S. Tesfatsion, Junjie Sun, and Leigh Tesfatsion. Dc optimal power flow formulation and solution using quadprogj. *Economics Department, Iowa State University*, 6014:1–36, 2006.

[19] Kazem Zare, Antonio J. Conejo, Miguel Carrion, and Mohsen Parsa Moghaddam. Multi-market energy procurement for a large consumer using a risk-aversion procedure. *Electric Power Systems Research*, 80(1):63 – 70, 2010. ISSN 0378-7796. doi: http://dx.doi.org/10.1016/j.epsr.2009.08.006. URL http://www.sciencedirect.com/science/article/pii/S0378779609001904.

[20] R.D. Zimmerman, C.E. Murillo-Sanchez, and R.J. Thomas. Matpower: Steady-state operations, planning, and analysis tools for power systems research and education. *Power Systems, IEEE Transactions on*, 26(1):12–19, Feb 2011. doi: 10.1109/TPWRS.2010.2051168.

5

Collaborative Model Predictive Control for a Distributed Market-Grid Coupling

5.1 ABSTRACT AND CONTEXT

In Chapter 4, we presented a MPC-based feedback control system for coupling a single grid unit with a correspondent local market. We formulated the local market-grid coupling as a single feedback control loop with a centralized MPC problem. In this chapter, we focus on a further coupling model for distributed power markets and power grids, for which we extend the proposed feedback control framework in Chapter 4 with a distributed MPC approach. Then, the centralized MPC problem for a single grid unit will be extended towards a distributed MPC problem for distributed grid units. The main contribution of this chapter is the mathematical model of a distributed closed-loop feedback control system for a collaborative power dispatch strategy among the given distributed grid units. By means of multi-agent systems (MAS), the distributed control architecture is implemented. To investigate the mutual influence between power markets and power grids based on the proposed collaborative control loops, numerical simulations are conducted with respect to both market price and load dispatch stability. The content of this chapter is based on the paper published at IEEE ENERGYCON 2016 [2].

5.2 DISTRIBUTED CONTROL SYSTEM DESIGN

In the previous chapter, we introduced a single MPC-based control loop that is designed for a single grid unit to regulate the stability of both market price and power dispatch. The control problem is formulated as a centralized MPC problem. On the one hand, for future large-scale power systems with a high penetration of distributed generation units and complex constraints, a globally centralized MPC technique can be very computationally intensive. On the other hand, Hatziargyriou et al. [4] and Menniti et al. [5] proposed a local power market within a microgrid to optimize its operational costs, as the local market process can not only benefit the end user with lower prices and better demand side management [5], but also support the transition to islanded operation mode in terms of security issues [4]. That means that distributed local power markets will bring distributed power grids both economic and operational advantages. Therefore, in this chapter, we focus on the extension of the centralized control loop towards collaborative control loops in a distributed manner, in order to develop a distributed market-grid coupling system. The collaborative aspect among the control loops is presented in terms of information exchange. The information exchange refers not only to an information sharing between the interconnected grid units but also an exchange of output values between individual grid units and correspondent local market units. Thus, the local optimization formulation for each individual grid unit is transformed to a collaborative optimization formulation, in which physically interconnected grid units can solve their own ODD problems with a collaborative capability for suppressing certain constraints' violation.

For modeling the above proposed distributed control loops, the underlying model of a closed-loop feedback control system that is presented in Chapter 4 remains unchanged. The model extension refers mainly to a distributed control strategy that is implemented in a distributed control architecture by the MAS approach. Before we introduce the distributed control strategy, the underlying control loop model is summarized as follows:

- A local feedback control loop depicted in Figure 25 is generalized for each individual grid unit;

- It consists of 8 modules;

- 1 **Grid** module describes the topology of the grid unit;

- 1 **OPF** module formulates the correspondent OPF problem of the given grid unit;

- 1 **Load Model** module calculates power loads and load forecasts on each bus node within the given grid unit;

- 1 **LMP** module determines nodal prices based on the **OPF** module;

- 1 **Market** module determines local market clearing prices based on the outputs of **Load Model** and **LMP**;

- 1 **MPC** module determines generation re-dispatch schedules;

- 1 **Control Variable** module represents generation set-points from the **MPC** module;

- 1 **MIQP** module represents the MIQP optimization solver for both **OPF** and **MPC**.

As formulated in Chapter 4, **OPF** and **MPC** optimization problems are control-relevant modules for market price and power dispatch, respectively. From the grid point of view, regardless of whether there is only one single grid unit or there are several interconnected grid units, the OPF-based market price control focuses on a stabilization of the local market price. This market price stabilization is rather a local optimization problem of the individual grid unit. However, the MPC-based power dispatch control is concerned with a stabilization of the power dispatch between all interconnected grid units. When the inter-transmission is taken into account, this power dispatch stabilization is then rather a global optimization problem of all interconnected grid units. Therefore, the following distributed control strategy is only applied in the **MPC** module.

5.2.1 *Distributed MPC Strategy*

As mentioned in the previous section, a centralized MPC approach for solving the ODD problem can be computationally intensive. A distributed (DMPC) approach is an efficient alternative to decompose the complex ODD problem into sub-problems with lower complexity. Then, individual sub-problems in terms of smaller optimization problems can be solved faster in parallel. However, due to incomplete information for the sub-problems, the distributed approach can determine locally optimal solutions but can not guarantee a global optimum for the original ODD problem based on the local optimal solutions.

The distributed MPC (DMPC) strategy that is employed in this dissertation is adopted to decompose the overall grid into interconnected grid units, thereby solving local ODD problems of individual grid units in consideration of inter-transmission power flows. Venkat et al. [8] introduced two different formulations of a distributed MPC strategy, i.e. *communication-based MPC* and *cooperation-based MPC*. In order to guarantee the closed-loop stability, the cooperation-based distributed MPC strategy is preferred, since the communication-based strategy focuses mainly on the exchange of sensory data (e.g. state and input variables) for each grid unit's MPC. With the cooperation-based strategy, each grid unit's MPC exchanges with its neighboring (meaning interconnected) grid units not only the information of state and input variables, but also the controller (optimization) objectives. Figure 41 shows a sketch of the proposed distributed control architecture, which demonstrates a cooperation-based MPC strategy for three interconnected grid units.

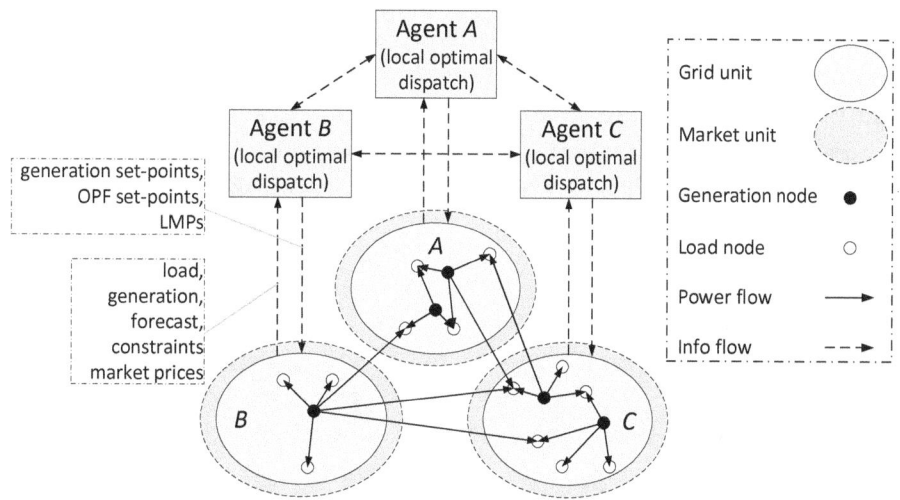

Figure 41.: Sketch of a DMPC architecture for a power system of three interconnected grid units with individual local market units. Within each grid unit, every solid cycle stands for a generation node, while every hollow cycle is a load node. The solid arrows refer to power flows (in transmission branches) between generators and consumers. Information flow and exchange between control agents and market-grid units are indicated by dashed arrows

As depicted in Figure 41, a global dispatch optimization problem is split into three local optimal dispatch problems that are solved by control agents of the correspondent grid units. In the figure, we generalize each control agent by **Agent**. The individual agents are distributed according to the partitioned grid units over the entire grid network. Each of them is locally responsible for both **OPF** and **MPC** optimization problems, namely dispatching and controlling the capacities of the generation and load units, and determining the nodal prices, within the correspondent grid unit. Each agent, once its local optimization problems are solved, sends its local objective solution to the neighboring agents, who are located within a predefined communication range R_c or represent interconnected grid units. Details about the agent model for implementing this DMPC architecture is provided in Section 5.4.

5.2.2 *Agent Cooperation Algorithm*

The optimization problems are solved iteratively based on an agent cooperation as illustrated in Figure 42. In this dissertation, communication rules for the agent cooperation are implemented with a parallel scheme [6]. That means that all agents solve local optimization problems at the same time. Also sending and receiving optimization results between agents may happen concurrently as long as optimization results become available. In comparison to a serial scheme [6], agents within a parallel implementation don't have to wait for updated information of neighboring agents until they can proceed the next optimization iteration.

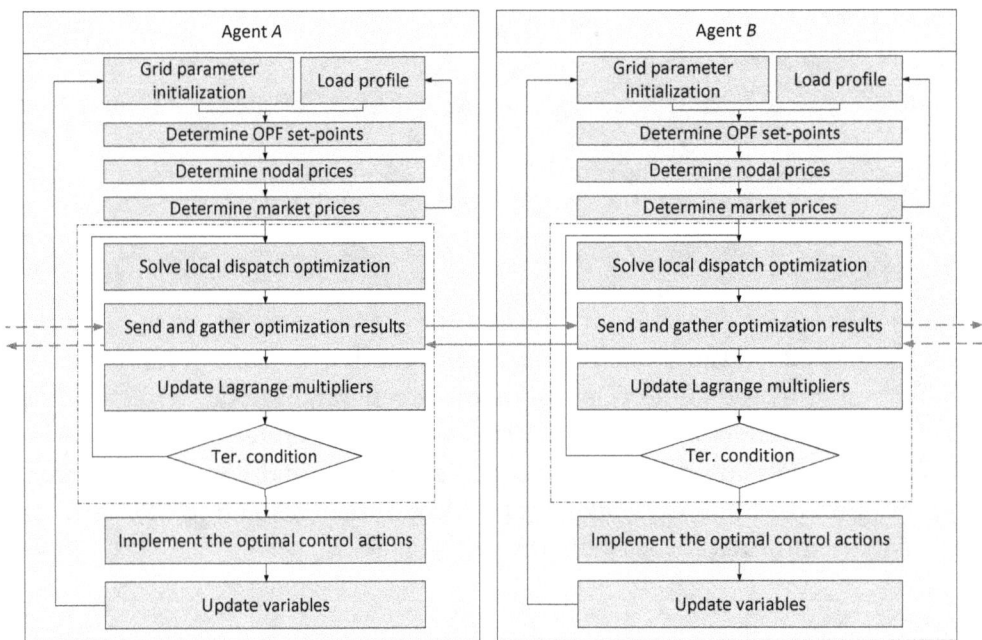

Figure 42.: Schema of the agent cooperation algorithm for both OPF and MPC optimization problems with a parallel implementation for two example neighboring agents; each agent is responsible for one grid unit; the red arrows represent the exchange of optimization results

All agents start in parallel to compute their local optimization. The already available optimization solutions of individual agents are taken into account by correspondent neighboring agents that are ready to start a next iteration for updating their optimization. In general, the parallel implementation of this agent cooperation algorithm that is schematized in Figure 42 can be summarized as follows beginning at dispatch time step k and optimization iteration s:

1. All agents start in parallel to load initial parameters. In the meantime, load forecasting for each grid unit at current dispatch time step k is conducted;

2. Subsequently, all agents determine their own local OPF set-points, nodal prices and market prices for the current time step;

3. Once the agent A (an exemplary agent represents each agent) solves its MPC optimization problem, it broadcasts the solution of iteration s to the neighboring agents who lie within the communication range R_c or represent interconnected grid units, such as the agent B (an exemplary neighboring agent);

4. After the agent B has finished solving its MPC optimization problem at iteration s, it updates its Lagrange multipliers and received information from all neighboring agents including the agent A;

5. Every agent moves to the next iteration $s + 1$, and the algorithm continues jumping to step 3), which means to solve its MPC optimization problem again, until the Lagrange multipliers do not change anymore;

6. Then, the algorithm continues with time step $k + 1$.

5.3 OPTIMIZATION PROBLEMS EXTENSION

In Chapter 4, we presented the formulation of two optimization problems, i.e. the **market price optimization** and the **power dispatch optimization**, which refer to the control objectives of both market and grid within the proposed local control loop of a single grid unit. As illustrated in Section 5.2.2, due to the locality of the **market price optimization**, the above proposed distributed control strategy for the extension of distributed and interconnected grid units focuses mainly on the **power dispatch optimization**. That means for the implementation of the agent cooperation algorithm in Section 5.2.2, the complete formulations of the **market price optimization** are adopted, including the **OPF formulation** in (22), the **LMP formulation** in (36) or (37) and the **price stabilization** function in (38) or (39).

A DMPC problem formulation for the **power dispatch optimization** in the context of distributed and interconnected grid units is extended based on the **MPC formulation** for a single grid unit ((43) − (53)). We adopt the state-space representation in ((43) and ((44) as a general description of nodal dynamics for each grid unit G^i with n_i bus nodes. In comparison to the previous chapter, the inter-transmission power flows in the context of interconnected grid units are now taken into account for the following optimization, which means $\bar{w}_{i,in}(k) > \bar{0}$ and

$\bar{w}_{i,out}(k) > \bar{0}$. As the output expression (44) shown that the influence of nodal generation and load adjustments on nodal power injections $\bar{P}_{i,in}$ is reflected in both intra-transmission power flows and inter-transmission power flows. Besides, regarding information flow, neighboring agents have to exchange optimization results subject to the state-space representation with the interconnections (see Figure 42). Therefore, in the context of DMPC for determining the control vector $\bar{u}_i$ of each grid unit, we separate the overall objective function into a local grid sub-function $\Phi_{i,local}(\tilde{\bar{x}}_i(k+1|s), \tilde{\bar{u}}_i(k|s), \tilde{\bar{y}}_i(k|s))$ and a set of grid interconnection sub-functions $\Phi_{i,inter|j}(\tilde{\bar{w}}_{i,in}(k|s), \tilde{\bar{w}}_{i,out}(k|s))$ with $j \in \Omega_{G^i}$ denoting each interconnection to and from G^i, where s behind the vertical bar indicates the current optimization iteration, the tilde over the vector variables indicates variables over the prediction horizon T, e.g. $\tilde{\bar{u}}_i(k|s) \equiv \left[(\bar{u}_i(k|s))^{tr}, \ldots, (\bar{u}_i(k+T-1|s))^{tr} \right]^{tr}$. The DMPC problem for each individual grid unit G^i with the objective of maximizing the total social welfare is then formulated as:

$$\max_{\tilde{\bar{v}}_i(k|s)} \quad \Phi_{i,local}(\tilde{\bar{x}}_i(k+1|s), \tilde{\bar{u}}_i(k|s), \tilde{\bar{y}}_i(k|s)) + \sum_{j \in \Omega_{G^i}} \Phi_{i,inter|j}(\tilde{\bar{w}}_{i,in}(k|s), \tilde{\bar{w}}_{i,out}(k|s))$$

(54)

$$\tilde{\bar{v}}_i(k|s) \equiv \{\tilde{\bar{x}}_i(k+1|s), \tilde{\bar{u}}_i(k|s), \tilde{\bar{y}}_i(k|s), \tilde{\bar{w}}_{i,in}(k|s), \tilde{\bar{w}}_{i,out}(k|s)\}$$

subject to the equations (43) – (44) and the following constraints:

$$\|\Delta \bar{s}_i(k)\|_1 = \|\Delta \bar{d}_i(k)\|_1 \tag{55}$$

$$\bar{r}_i^{down} \leq \Delta \bar{s}_i(k) \leq \bar{r}_i^{up} \tag{56}$$

$$\bar{s}_i^{min} \leq \bar{s}_i(k) \leq \bar{s}_i^{max} \tag{57}$$

$$\bar{P}_{i,in}^{min} \leq \bar{P}_{i,in}(k) \leq \bar{P}_{i,in}^{max} \tag{58}$$

$$\bar{w}_{i,in}^{min} \leq \bar{w}_{i,in}(k) \leq \bar{w}_{i,in}^{max} \tag{59}$$

$$\bar{w}_{i,out}^{min} \leq \bar{w}_{i,out}(k) \leq \bar{w}_{i,out}^{max} \tag{60}$$

$$\bar{w}_{i \leftarrow j,in} = \bar{w}_{j \rightarrow i,out}, \quad j \in \Omega_{G^i} \tag{61}$$

$$\bar{w}_{i \rightarrow j,out} = \bar{w}_{j \leftarrow i,in}, \quad j \in \Omega_{G^i} \tag{62}$$

Equality constraint (55) balances total generation adjustments with total load changes. Inequality constraint (56) denotes the down and up ramp rates of generators. Inequality constraint (57) indicates upper and lower bounds of the nodal generation. Inequality constraints (58), (59) and (60) indicate the intra- and inter-transmission power flow limits. The last two equality constraints (61) and (62) imply that the inter-transmission power flow input to the grid unit G^i from G^j must be equal to the inter-transmission power flow output from G^j to G^i, and vice versa. Under these constraints, the optimization problem is solved iteratively by means of Lagrange multipliers at each dispatch time step k as illustrated in Section 5.2.2, in order to obtain the convergent solution until Lagrange multipliers can not be updated any more.

In order to determine the control vector $\bar{u}_i(k|s)$ based on the above DMPC formulation, we introduce social welfare expressions for both the local term and interconnection term into the DMPC objective function in (54).

$$
\begin{aligned}
&\Phi_{i,local}\left(\tilde{\bar{x}}_i(k+1|s), \tilde{\bar{u}}_i(k|s), \tilde{\bar{y}}_i(k|s)\right) \\
&= \sum_{l=0}^{T-1} \left(\bar{\pi}_i(k+l)\right)^{tr} \cdot
\begin{bmatrix} \bar{x}_i(k+l+1|s) \\ \bar{u}_i(k+l|s) \\ \bar{y}_i(k+l|s) \end{bmatrix} \\
&\quad - \begin{bmatrix} \bar{x}_i(k+l+1|s) \\ \bar{u}_i(k+l|s) \\ \bar{y}_i(k+l|s) \end{bmatrix}^{tr} \cdot Q_i \cdot
\begin{bmatrix} \bar{x}_i(k+l+1|s) \\ \bar{u}_i(k+l|s) \\ \bar{y}_i(k+l|s) \end{bmatrix} \\
&\quad - R_i^{tr} \cdot \begin{bmatrix} \bar{x}_i(k+l+1|s) \\ \bar{u}_i(k+l|s) \\ \bar{y}_i(k+l|s) \end{bmatrix} \\
&= \sum_{l=0}^{T-1} \left(\left(\bar{\pi}_i(k+l)\right)^{tr} - R_i^{tr} \right) \cdot
\begin{bmatrix} \bar{x}_i(k+l+1|s) \\ \bar{u}_i(k+l|s) \\ \bar{y}_i(k+l|s) \end{bmatrix} \\
&\quad - \begin{bmatrix} \bar{x}_i(k+l+1|s) \\ \bar{u}_i(k+l|s) \\ \bar{y}_i(k+l|s) \end{bmatrix}^{tr} \cdot Q_i \cdot
\begin{bmatrix} \bar{x}_i(k+l+1|s) \\ \bar{u}_i(k+l|s) \\ \bar{y}_i(k+l|s) \end{bmatrix}
\end{aligned}
$$

$$(63)$$

$$(64)$$

This local term is adopted from the MPC objective function of a single local grid unit in Chapter 4.

$$
\begin{aligned}
&\Phi_{i,inter|j}\left(\tilde{\bar{w}}_{i,in}(k|s), \tilde{\bar{w}}_{i,out}(k|s)\right) \\
&= \left[\left(\tilde{\bar{\pi}}_{i\to j}(k)\right)^{tr} \quad \left(-\tilde{\bar{\pi}}_{i\leftarrow j}(k)\right)^{tr} \right] \cdot
\begin{bmatrix} \tilde{\bar{w}}_{i\to j,out}(k|s) \\ \tilde{\bar{w}}_{i\leftarrow j,in}(k|s) \end{bmatrix} \\
&\quad - \left[\left(\tilde{\bar{\lambda}}_{i\to j}(k|s)\right)^{tr} \quad \left(-\tilde{\bar{\lambda}}_{i\leftarrow j}(k|s)\right)^{tr} \right] \cdot
\begin{bmatrix} \tilde{\bar{w}}_{i\to j,out}(k|s) \\ \tilde{\bar{w}}_{i\leftarrow j,in}(k|s) \end{bmatrix} \\
&\quad - \frac{c}{2} \left\| \begin{bmatrix} \tilde{\bar{w}}_{j\to i,out}(k|s-1) - \tilde{\bar{w}}_{i\leftarrow j,in}(k|s) \\ \tilde{\bar{w}}_{j\leftarrow i,in}(k|s-1) - \tilde{\bar{w}}_{i\to j,out}(k|s) \end{bmatrix} \right\|^2 \\
&\quad - \frac{b-c}{2} \left\| \begin{bmatrix} \tilde{\bar{w}}_{i\to j,out}(k|s) - \tilde{\bar{w}}_{i\to j,out}(k|s-1) \\ \tilde{\bar{w}}_{i\leftarrow j,in}(k|s) - \tilde{\bar{w}}_{i\leftarrow j,in}(k|s-1) \end{bmatrix} \right\|^2
\end{aligned}
$$

$$(65)$$

where in the interconnection term, $\bar{\pi}_{i\to j}(k)$ and $\bar{\pi}_{i\leftarrow j}(k)$ denote local market prices of the grid unit G^i and G^j, respectively. $\bar{\lambda}_{i\to j}(k|s)$ and $\bar{\lambda}_{i\leftarrow j}(k|s)$ are Lagrange multipliers of the equality constraints (61) and (62). Both positive scalars c and b are defined as penalization constants (the definition is detailed in [6]), where c penalizes the imbalance of the inter-transmission power flows between two interconnected grid units and $b - c$ penalizes the deviation of the inter-transmission power flows of one grid unit between the current optimization iteration s and the previous one $s - 1$. Moreover, the update of Lagrange multipliers at each new iteration is conducted as follows:

$$\tilde{\tilde{\lambda}}_{i \to j}(k|s+1) = \tilde{\tilde{\lambda}}_{i \to j}(k|s) + c \cdot \left(\tilde{\tilde{w}}_{j \leftarrow i,in}(k|s+1) - \tilde{w}_{i \to j,out}(k|s+1) \right) \tag{66}$$

$$\tilde{\tilde{\lambda}}_{i \leftarrow j}(k|s+1) = \tilde{\tilde{\lambda}}_{i \leftarrow j}(k|s) + c \cdot \left(\tilde{\tilde{w}}_{i \leftarrow j,in}(k|s+1) - \tilde{w}_{j \to i,out}(k|s+1) \right) \tag{67}$$

5.4 MULTI-AGENT SYSTEM

For the implementation of the above DMPC problem, we model a decentralized MAS to realize the agent cooperation algorithm described in Section 5.2.2. Considering the dynamic behaviors of each grid and market units, and intra-grid and inter-grid control requirements, we propose a hierarchical MAS architecture with 6 agent types to execute the local market optimization, the intra-grid optimization and the inter-grid optimization.

Following the hierarchical MAS concept of the previous research work [1, 9, 7], the main idea of a hierarchical MAS application for our DMPC problem is to split the complex control problem of the entire grid network into several small local optimization problems of individual grid units. Each local optimization problem can be solved by a grid agent that has a certain degree of autonomy for making decisions based on the received information from other grid agents and component agents of its correspondent grid unit. The proposed MAS consists of three hierarchy levels, i.e. **Upper Level**, **Middle Level** and **Lower Level**, and 6 agent types, i.e. **DMPC Agent**, **Grid Agent**, **Bus Agent**, **Branch Agent**, **Load Agent** and **Generation Agent**. As depicted in Figure 43, the **DMPC Agent** is located at the **Upper Level**, which presents the DMPC problem formulation and represents a correspondent optimization solver that needs the information from grid agents of correspondent grid units; The **Grid Agent** is located at the **Middle Level**, which is responsible for three local optimization problems, i.e. the **OPF** problem, the local **Market** optimization problem and the local **MPC** problem, and represents as well a optimization solver that needs the information from agents of the grid components; The 4 grid component agents, namely the **Bus Agent**, the **Branch Agent**, the **Load Agent** and the **Generation Agent** are located at the **Lower Level**, which present the static and dynamic information of the grid components. In this MAS architecture, all the above agents work in a decentralized and/or coordinated control manner and their characteristics are specified as follows.

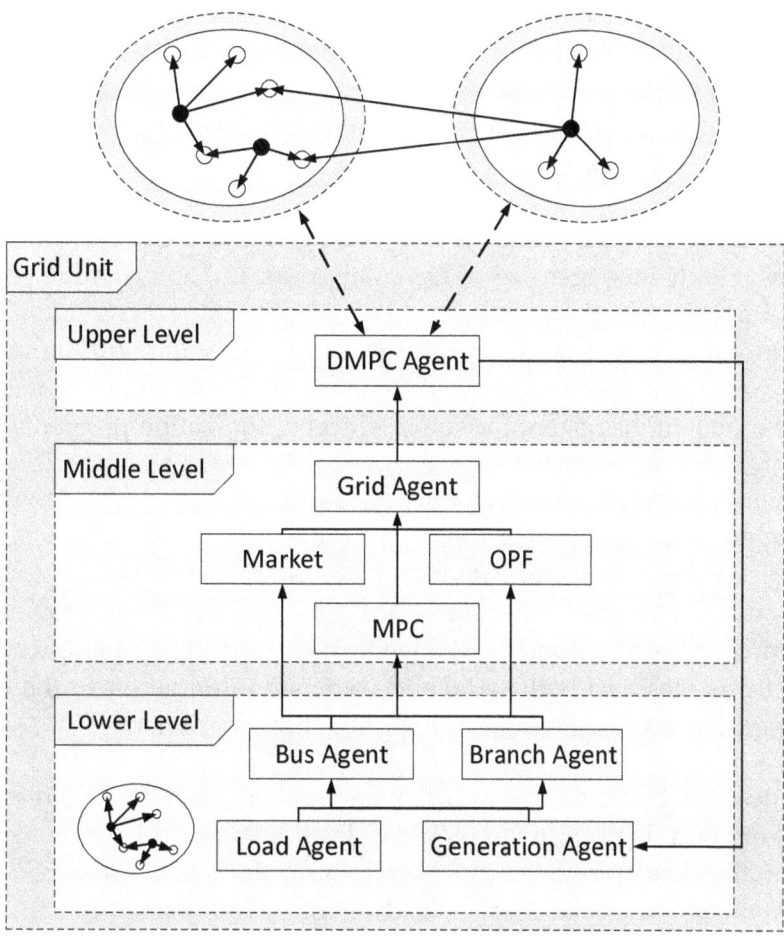

Figure 43.: Architecture of the proposed MAS

DMPC Agent: Each grid unit has a DMPC agent that is the core agent for solving the DMPC problem in (54). A DMPC agent optimizes the local power dispatch with the information of state and input variables as well as the controller (optimization) results of its neighboring DMPC agents. It receives local grid information from own grid agent and its neighboring grid agents, and exchanges the controller results iteratively with neighboring DMPC agents. After optimization at each time step, the control values in terms of generation set-points are fed back by its DMPC agent to the correspondent generation agents.

Grid Agent: Each grid unit has a grid agent that is responsible for 3 local optimization problems, which determine and stabilize local market prices and local power dispatch based on the OPF formulation in (22), the LMP formulation in (36) or (37), the price stabilization function in (38) or (39) and the local term of the DMPC objective function in (51). A grid agent represents a grid unit to communicate with its neighboring grid agents, and provides local optimization results to its DMPC agent.

Bus Agent: Each grid unit has a list of bus agents that represent physical bus nodes. Each bus agent records nodal constraint information, e.g. nodal admittance, generation capacity, load limit, voltage magnitude limit, etc., and monitors real-time generation and load information through generation and load agents. Both static and dynamic information are provided to the correspondent grid agent for solving local optimization problems.

Branch Agent: Each grid unit has a list of branch agents that represent physical branch models for a connection between bus nodes and a physical power delivery. Each branch agent records transmission constraints, i.e. transmission capacity and branch admittance, and calculates real-time power injection based on generation and load agents. The same as a bus agent, a branch agent provides also both static and dynamic information to the correspondent grid agent for solving local optimization problems.

Load Agent: Each bus agent has a correspondent load agent that represents power consumption (load) at that bus node. Each load agent has a load model for generating load profiles and forecasts at an initialization step. After optimization at each time step, a load agent is responsible for an update in terms of a load increase or a load shedding determined by the correspondent DMPC agent.

Generation Agent: Each bus agent has a correspondent generation agent that represents power generation at that bus node. Each generation agent records nodal generator information, e.g. generator type, ramping constraints, etc., and monitors real-time power generation output of correspondent generators. After optimization at each time step, a generation agent controls its generators with updated generation set-points determined by the correspondent DMPC agent.

Furthermore, a class diagram for the implementation of the DMPC problem in Section 5.3 is depicted in Figure 72 in Appendix. All the defined 6 agents are demonstrated as agent classes in the class diagram. As shown, the `GridAgent` is responsible for the local optimizations, while the `DMPCAgent` is responsible for the global optimization through a communication with the neighboring `DMPCAgent`.

5.5 TEST AND EVALUATION

In this section, we re-introduce an evaluation scenario based on the IEEE 300 bus test case that has already been discussed in Section 4.5 in Chapter 4 briefly. We re-iterate the simulation-based numerical results to demonstrate the capability of the proposed collaborative MPC feedback loops. We divide the 300-bus power network into 4 zones, where bus nodes are electrically close to each other within each zone with respect to power losses. Thus, the 300-bus power grid is disaggregated into 4 grid units (G^1, G^2, G^3 and G^4) based on the grid disaggregation algorithm presented in Chapter 3 and all bus nodes with branches are split up into correspondent grid units, see Table 8. As described in Chapter 3, each grid network or grid unit in this dissertation is modeled as a directed graph, so that the interconnection between grid units in Table 8 refers only to a unidirectional power transmission.

Table 8.: The number of bus nodes, branches and interconnections of individual grid units

Grid Units	# Generator Bus	# Load Bus	# Branch	# Interconnection to			
				G^1	G^2	G^3	G^4
G^1	26	96	166	-	3	6	-
G^2	22	58	114	-	-	-	-
G^3	16	47	83	1	-	-	-
G^4	5	30	37	-	-	-	-

Similar to the previous chapter, for each generator bus, we assume that its aggregate cost is estimated by a quadratic cost function $c_i(x) = \alpha_2^i \cdot x^2 + \alpha_1^i \cdot x + \alpha_0^i$. $\alpha_{0,1,2}^i$ are cost coefficients for polynomial cost functions from the test case and $\alpha_2^i > 0$ ensures the convexity. We further assume that for each load bus, the consumers' total utility value is expressed by a logarithmic utility function $u_j(x) = \beta^j \cdot \ln x$, so that the inverse function $d_j(\pi_j)$ of power load can be expressed as $d_j = \beta^j / \pi_j$, where π_j indicates nodal local market prices. The logarithmic function implies normal consumers' risk-averse behavior in microeconomics, which means that most of the consumers prefer a fixed budget for their energy bills [3, 10]. Furthermore, different values of the coefficient β^j represent different consumer/load types. For simplicity in our test, we simulate all load buses with the same consumer/load type, i.e. $\forall j : \beta^j = 1500$. The same as in the evaluation configuration of Chapter 4, for the pricing function, we employ the λ-based method expressed in (38), where $\gamma = 0.05$.

Both optimizations in each local control loop are conducted by a MIQP (mixed-integer quadratic programming) solver in Matlab. To make a k-step simulation, we need the load/consumption information of those k steps. For more realistic nodal loads with stochastic aspects, we simulate nodal power loads $d_j(k)$ with an additive white noise $\epsilon_j \sim \mathcal{N}\left(0, \sigma_j^2\right)$ and cap them with lower and upper bounds (d_{min} and d_{max}): $d_j(k) = \max(d_{min}, \min(d_{max}, \beta^j / \pi_j(k) \cdot (1 + \epsilon_j)))$, where $d_{min} = 0$, d_{max} is assigned by the power demand from the test case, and $\sigma_j = 1/3$.

First, we present the optimization results of these 4 individual grid units without a utilization of the inter-transmission, which means that all 4 grid units are individually optimized based on own local control loop without the DMPC agent. The simulation results with 200 time steps for nodal load adjustments Δd are shown in Figure 44, for which the prediction horizon of the MPC was set with 3 options, i.e. $T = 1$ (left column), $T = 5$ (middle column) and $T = 10$ (right column). As depicted in the three bottom sub-figures that refer to the simulation results of the grid unit G^4, the load dispatch Δd has in each case either strong oscillations or extreme peaks, which implies no good stability regarding the local load dispatch optimization. For the other 3 grid units, we notice only small changes regarding the load dispatch Δd in all 3 prediction horizon cases except for the beginning phase that refers to a MPC initialization.

Moreover, the local market price π of G^1, G^2 and G^3 depicted in Figure 45a, 45b and 45c, shows in all 3 prediction horizon cases a clear convergence. This in turn confirms the defined stability of the proposed local market-grid coupling that the MPC-based feedback control system can maintain the steady-state operation of individual local markets. However, the local market price π of G^4 in Figure 45d exhibits in each prediction horizon case strong oscillation and no clear convergence. Moreover, except for the grid unit G^4, a comparison of the results of all three prediction horizon cases shows that both the load dispatch Δd and the local market price π don't significantly differentiate.

At this point, we analyze the execution time of solving the optimization problems in the local control loop of each grid unit. Since the simulation runs of 4 grid units are conducted in parallel, Figure 46 shows the maximum run time data of the 200 simulation time steps for all 3 prediction horizon cases, i.e. $T = 1$, $T = 5$ and $T = 10$. We notice that the average maximum run time of each simulation time step is about 40 s in case $T = 1$, 50 s in case $T = 5$ and 60 s in case $T = 10$. There are some run time peaks in case $T = 10$, which reach about 200 s, the execution time in each case is nevertheless averagely less than 200 s, which answers our question left in Chapter 3 that the MPC-based control problem can be solved within the same time scale as the set requirement of the proposed market-grid coupling.

Then, as the next step, we show the optimization results of these 4 interconnected grid units, which means that all 4 grid units are optimized based on the collaborative control loops with the DMPC agents. The simulation results with

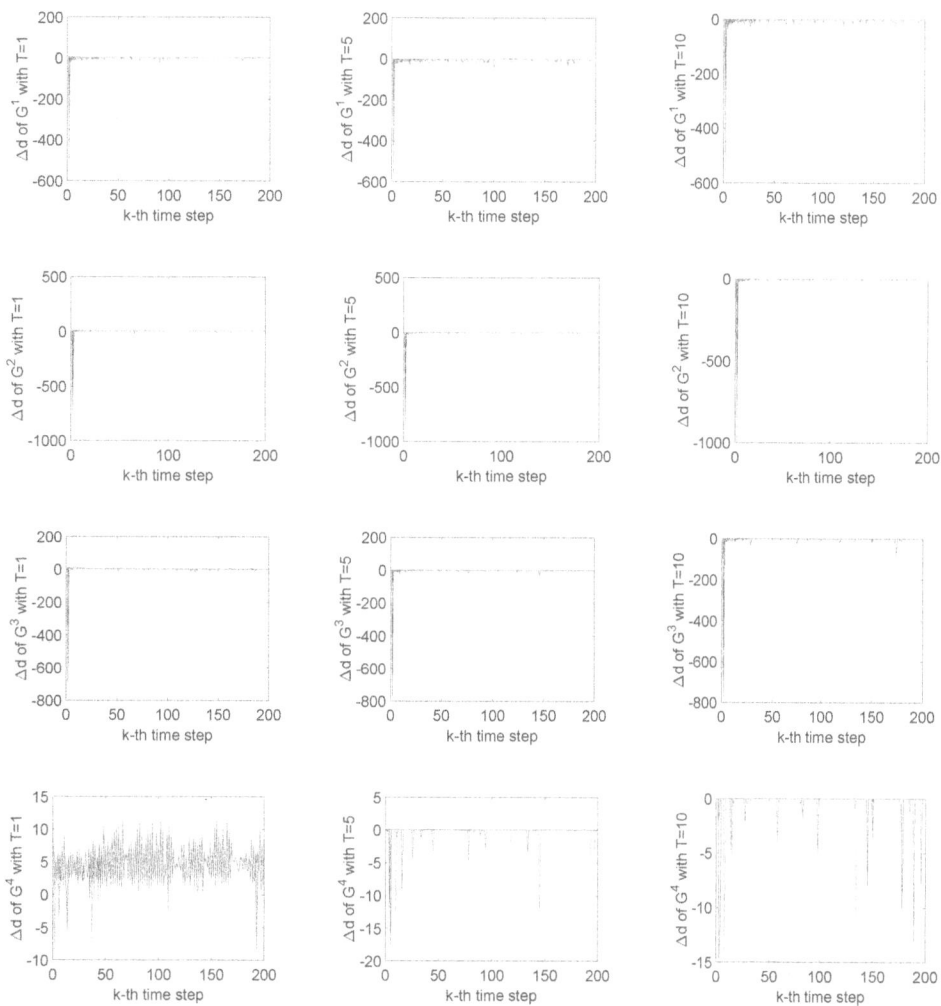

Figure 44.: Evolution of nodal load adjustments Δd with $T = 1$ (left column), $T = 5$ (middle column), and $T = 10$ (right column) of all 4 grid units without any agent cooperation

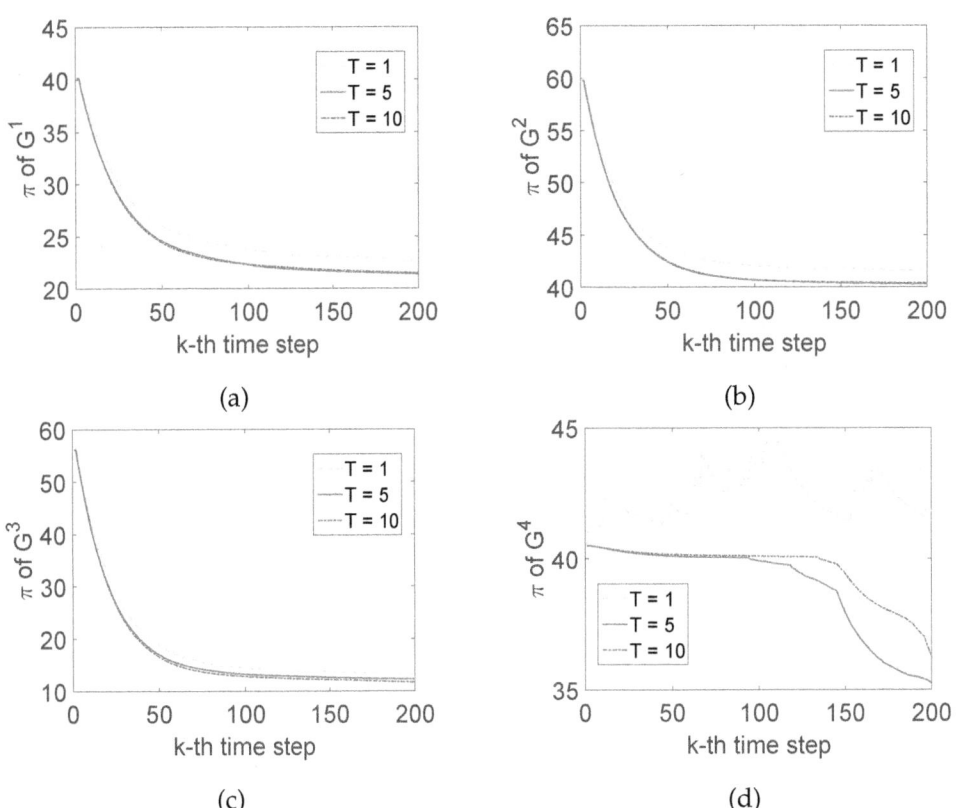

Figure 45.: Evolution of local market prices π with $T = 1, 5, 10$ of all 4 grid units without any agent cooperation: G^1 ((a) upper left), G^2 ((b) upper right), G^3 ((c) bottom left) and G^4 ((d) bottom right)

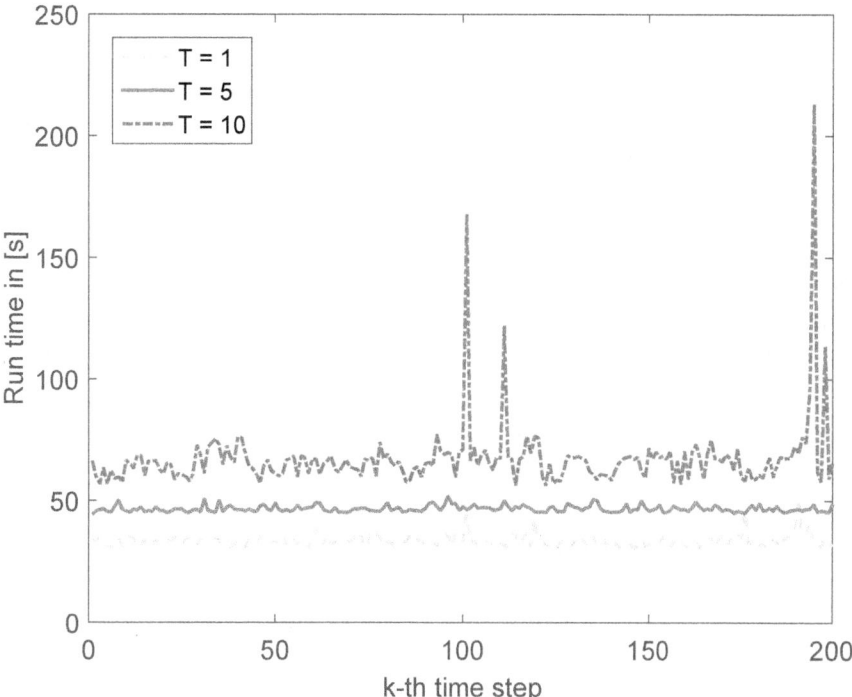

Figure 46.: Evolution of simulation run time with $T = 1, 5, 10$ on all 4 grid units in parallel without any agent cooperation

200 time steps for nodal load adjustments Δd are shown in Figure 47, for which the prediction horizon of the MPC was set with 3 options, i.e. $T = 1$ (left column), $T = 5$ (middle column) and $T = 10$ (right column). At each simulation step, it is guaranteed that the OPF for all 4 grid units converges. As shown in Table 8, the grid unit G^4 is still an isolated unit. Therefore, similar to the previous case without any DMPC agent, as depicted in the three bottom sub-figures that refer to the simulation results of the grid unit G^4, there is no possibility for G^4 to stabilize its load dispatch Δd through no inter-transmission. For the other 3 grid units, due to the interconnections between G^1, G^2 and G^3, we notice a clear convergence with only small changes regarding the load dispatch Δd in all 3 prediction horizon cases except for the beginning phase that refers to a MPC initialization.

Similar to the previous case without any agent cooperation, the local market price π of G^1, G^2 and G^3 under an agent cooperation as depicted in Figure 48a, 48b and 48c, shows in all 3 prediction horizon cases a clear convergence. This confirms the defined stability of the proposed local market-grid coupling that the MPC-based feedback control system can maintain the steady-state operation of individual local markets. Moreover, from a comparison between the local market price without any agent cooperation in Figure 45 and with an agent cooperation in Figure 48, we notice that the stabilized price π determined with an agent cooperation for these interconnected 3 grid units scales a larger price range than without any agent cooperation. That means that the interconnected grid units can benefit from this collaborative optimization not only for a stable load dispatch but also for a price advantage. However, due to the isolated situation, the local market price π of G^4 cannot benefit from this collaborative optimization. As shown in Figure 48d, its local market price exhibits in each prediction horizon case strong oscillation and no clear convergence. In addition, a comparison of the results of all 3 prediction horizon cases shows that the local market price π can be affected significantly by different prediction horizon values with respect to both magnitude and stability. Overall, the local market price π are stabilized with less oscillation at $T = 5, 10$ than at $T = 1$, which means that the more future information the DMPC agent perceives, the better stability the optimization results achieve.

In Chapter 3, we introduced that the time response requirement of both the intra-minute market and the power flow optimization is designed with a time scale of minutes. Therefore, similar to the previous case without any agent cooperation, we analyze also the execution time of solving the optimization problems based on the collaborative control loops with the DMPC agents. Figure 49 shows the run time data of the 200 simulation time steps for all 3 prediction horizon cases, i.e. $T = 1$, $T = 5$ and $T = 10$. We notice that the average run time of each simulation time step is about 45 s in case $T = 1$, 50 s in case $T = 5$ and 60 s in case $T = 10$, which is comparable with the case of individual local optimizations

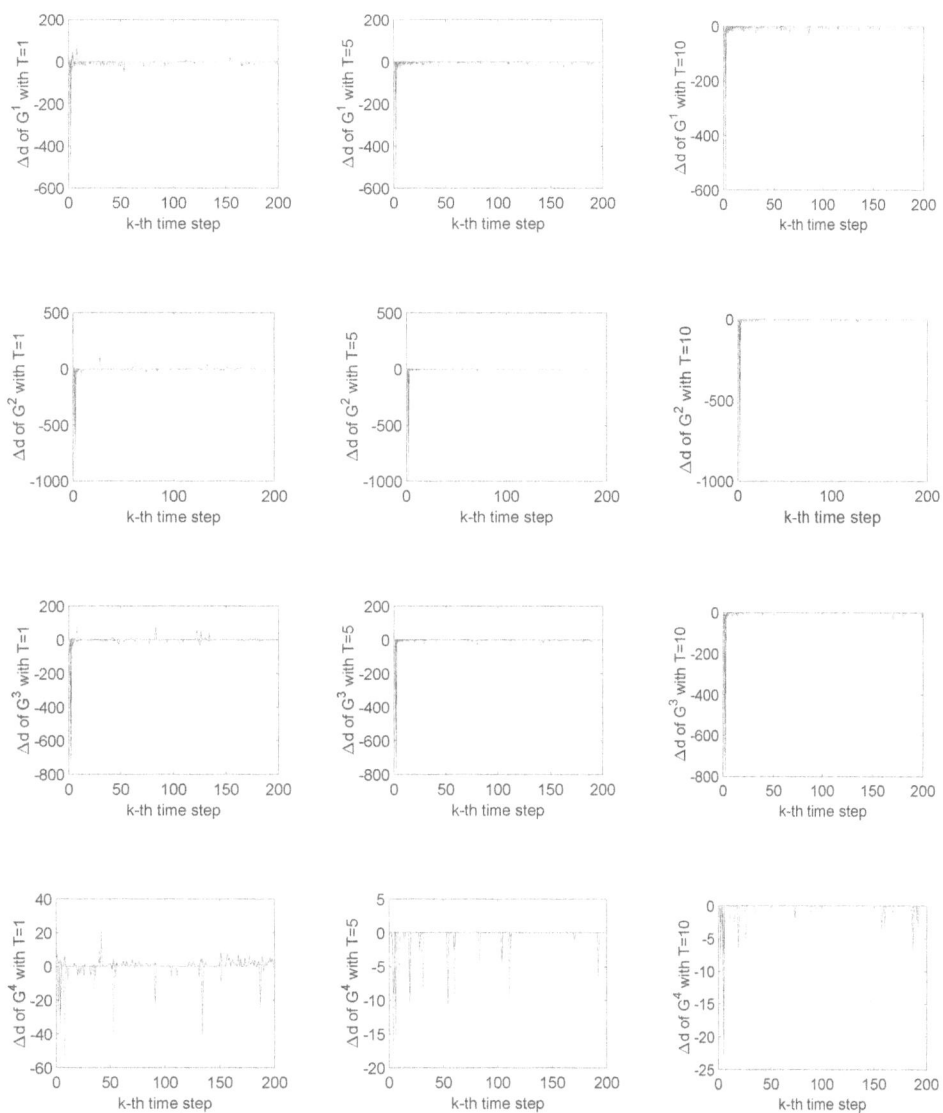

Figure 47.: Evolution of nodal load adjustments Δd with $T = 1$ (left column), $T = 5$ (middle column), and $T = 10$ (right column) of all 4 grid units with an agent cooperation

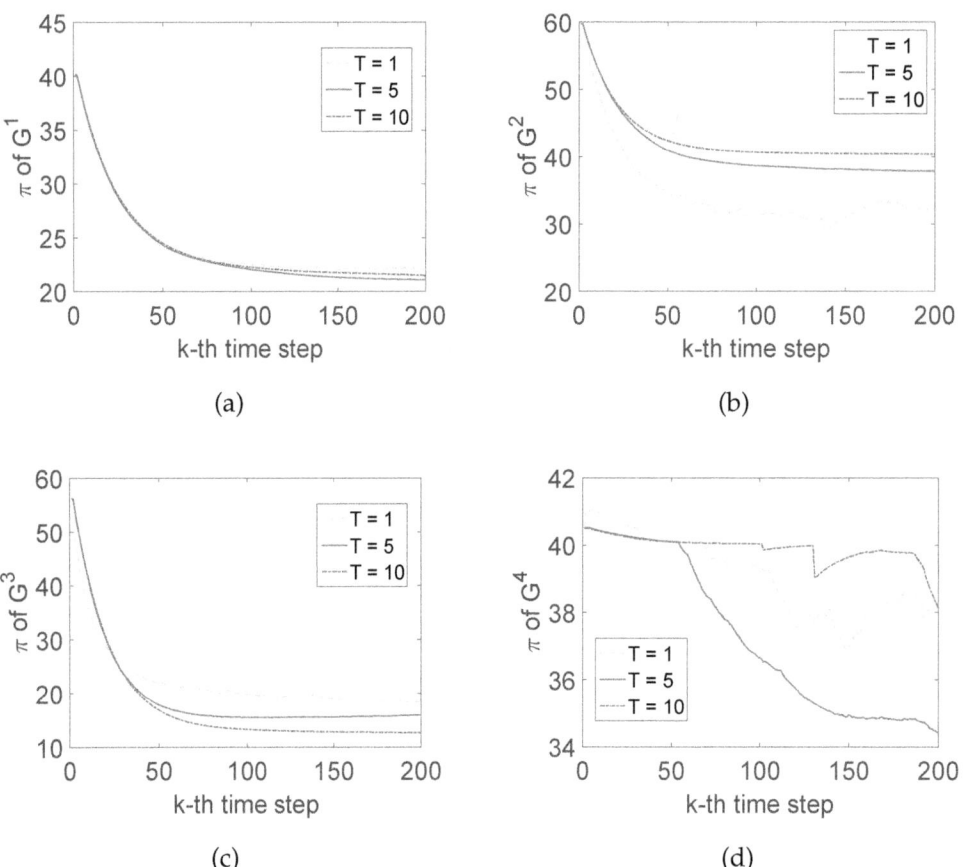

Figure 48.: Evolution of local market prices π with $T = 1, 5, 10$ of all 4 grid units with an agent cooperation: G^1 ((a) upper left), G^2 ((b) upper right), G^3 ((c) bottom left) and G^4 ((d) bottom right)

without the DMPC agent in Figure 46. Overall, in each case, the execution time of the collaborative control loops with the DMPC agents can be expected less than 70 s in average, which meets the time response requirement of the proposed market-grid coupling.

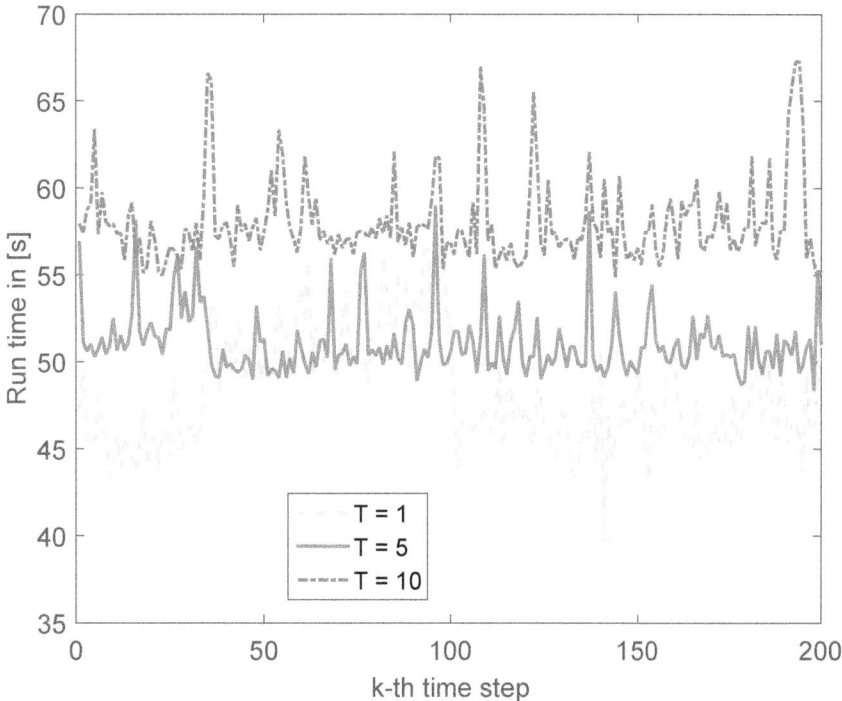

Figure 49.: Evolution of simulation run time with $T = 1, 5, 10$ on all 4 grid units with an agent cooperation

5.6 CONCLUSION

In this chapter, we extended the control framework in the previous chapter with a distributed MPC approach. By means of multi-agent systems (MAS), we implemented a distributed control architecture to realize the cooperation-based distributed MPC strategy, so that the centralized MPC problem for a single grid unit has been extended to a distributed MPC problem for distributed grid units. Within this distributed control architecture, we focused not only on an information sharing between interconnected grid units, but also on an output exchange between local grid units and correspondent local markets. After the presentation of the extended mathematical model for a collaborative power dispatch strategy, we introduced an evaluation use case based on the IEEE 300 bus test case, which has been modified based on our grid disaggregation algorithm to 4 case files representing 4 grid units, i.e. G^1, G^2, G^3 and G^4. Simulations for the distributed MPC problem were conducted in two different cases: (1) each grid unit was individually optimized based on own local control loop without the DMPC agent and (2) interconnected grid units are optimized based on the collaborative control loops with the DMPC agents. The simulation results showed that the interconnected grid units can benefit from the proposed collaborative MPC feedback loops for stabilizing their market price and power load dispatch. Overall, a comparison of different prediction horizon values indicated that the more future information the DMPC agent perceives, the more stability the optimization results achieve. Furthermore, the simulation run time confirmed that the proposed DMPC problem for each time step can be solved with a time scale of minutes. These conclude our proposition that a MPC-based feedback control system can couple the market with the grid as two control problems in a control loop as proposed, and an optimal dynamic dispatch stabilizes the market price of individual grid units.

BIBLIOGRAPHY

[1] Niannian Cai, Xufeng Xu, and J. Mitra. A hierarchical multi-agent control scheme for a black start-capable microgrid. In *Power and Energy Society General Meeting, 2011 IEEE*, pages 1–7, July 2011. doi: 10.1109/PES.2011.6039570.

[2] Yong Ding, Erwin Stamm, and Michael Beigl. Using model predictive control to interact distributed power markets and grids. In *in Proceedings of IEEE International Energy Conference (ENERGYCON 2016)*, Leuven, Belgium, April 2016. IEEE.

[3] Mehdi Farsi. Risk aversion and willingness to pay for energy efficient systems in rental apartments. *Energy Policy*, 38(6):3078–3088, 2010.

[4] ND Hatziargyriou, A Dimeas, AG Tsikalakis, JA Pecas Lopes, G Karniotakis, and J Oyarzabal. Management of microgrids in market environment. In *Future Power Systems, 2005 International Conference on*, pages 7–pp. IEEE, 2005.

[5] D. Menniti, Anna Pinnarelli, and N. Sorrentino. Operation of decentralized electricity markets in microgrids. In *Electricity Distribution - Part 1, 2009. CIRED 2009. 20th International Conference and Exhibition on*, pages 1–4, 2009.

[6] R.R. Negenborn, B. De Schutter, and H. Hellendoorn. Multi-agent model predictive control of transportation networks. In *Networking, Sensing and Control, 2006. ICNSC '06. Proceedings of the 2006 IEEE International Conference on*, pages 296–301, 2006. doi: 10.1109/ICNSC.2006.1673161.

[7] Fenghui Ren, Minjie Zhang, and D. Sutanto. A multi-agent solution to distribution system management by considering distributed generators. *Power Systems, IEEE Transactions on*, 28(2):1442–1451, May 2013. ISSN 0885-8950. doi: 10.1109/TPWRS.2012.2223490.

[8] A.N. Venkat, I.A. Hiskens, J.B. Rawlings, and S.J. Wright. Distributed mpc strategies with application to power system automatic generation control. *Control Systems Technology, IEEE Transactions on*, 16(6):1192–1206, Nov 2008. doi: 10.1109/TCST.2008.919414.

[9] Chun xia Dou and Bin Liu. Multi-agent based hierarchical hybrid control for smart microgrid. *Smart Grid, IEEE Transactions on*, 4(2):771–778, June 2013. ISSN 1949-3053. doi: 10.1109/TSG.2012.2230197.

Bibliography

[10] Kazem Zare, Antonio J. Conejo, Miguel Carrion, and Mohsen Parsa Moghaddam. Multi-market energy procurement for a large consumer using a risk-aversion procedure. *Electric Power Systems Research*, 80(1):63 – 70, 2010. ISSN 0378-7796. doi: http://dx.doi.org/10.1016/j.epsr.2009.08. 006. URL http://www.sciencedirect.com/science/article/pii/ S0378779609001904.

6

An Adaptive Load Forecasting Framework

6.1 ABSTRACT AND CONTEXT

In this chapter, we present a framework for implementing short-term load fore-casting, in which statistical time series prediction methods and machine learning-based regression methods, can be configured to benchmark their performance against each other on given datasets of power consumption and other related ex-ogenous variables. In addition to prediction methods, the performance of a load forecasting depends also on the quality of training data. This aspect of data quality is addressed by two characteristics of the framework regarding data collection and preprocessing. The first characteristic is to introduce a human activity variable as an additional load influencing factor which reflects anomalous load patterns by aperiodic human activity that can be recognized by a sensing infrastructure. The second characteristic is to wavelet transform training data during the preprocess-ing stage to better extract redundant information from consumption data. To investigate the capability of the proposed framework, 3 case studies are presented for predicting power consumption at different scales with different prediction lead times. The results indicate that, in general, the aggregation level of consumption data and activity data matters. Finally, we demonstrate a numerical simulation for the integration of this forecasting framework in the MPC problem formulation of Chapter 5. The simulation results show that by incorporating accurate load forecasting results into the MPC problem formulation, a further cost reduction and stabilization for the power re-dispatch can be achieved. The content of this chapter is based on a paper published at ACM AIIP 2013 [6], a paper published at ACM WoT 2014 [7] and two papers published at IEEE ISC2 2015 [9, 8].

6.2 INTRODUCTION AND BACKGROUND

Power load forecasting has significant influences on power system planning and operation, in particular short-term load forecasting (STLF) [20]. Many operational decisions such as generation scheduling, load management and system security assessment are made based on short-term forecasts. An accurate STLF model is required to relax the conflict between supply and demand for a power system [17] and provide anticipatory possibilities of load modifications for a demand side management (DSM) program [5, 13]. STLF refers to load forecasts of electric loads with lead times ranging from a few minutes to seven days ahead. The objective of STLF is to predict future power consumption based on historical consumption data and other exogenous variables, which works at different power aggregation levels. Both research work [1, 18, 10, 22] and competitions (e.g. Eunite 2001 competition and the load forecasting track of GEFCom2012) have shown that a system-level load forecasting that is used to predict the total load at e.g. bus or substation levels, can achieve good predictions due to repetitive load patterns. However, at household's or appliance level, the closeness to the end consumer implies that load at this level strongly features aperiodicity caused by the uncertainty of human behaviors and activities [11, 16].

Human activity is an important contributor to local power consumption, particularly in urban areas [19, 12, 14]. Among all top-down and bottom-up approaches of modeling the end-use power consumption, lots of recent research works [23, 25, 11, 16, 3, 15] have pointed out that the end-use power consumption is aperiodic and highly dependent on human habits, behaviors and activities, either the high-level social activity [17] or the appliance-level usage activity [11]. The question arises whether we could efficiently utilize the human activity information to enhance STLF models with respect to the forecasting accuracy.

An important problem in STLF is to select relevant variables and features based on a given training dataset, thereby including them appropriately in STLF models. In order to explore the influence of human activity on load forecasting, we propose to incorporate the human activity information of different scales as an exogenous influencing factor into STLF models. First, we need to identify human activities, like behaviors, situations and events on different scales, which have a relevant impact on short-term power load, and quantify their impact scales. Then, our work focuses on a correlation study for the STLF enhancement as depicted in Figure 50, between:

- Low-level activity information and low-level load information

- Low-level activity information and high-level load information

- High-level activity information and low-level load information

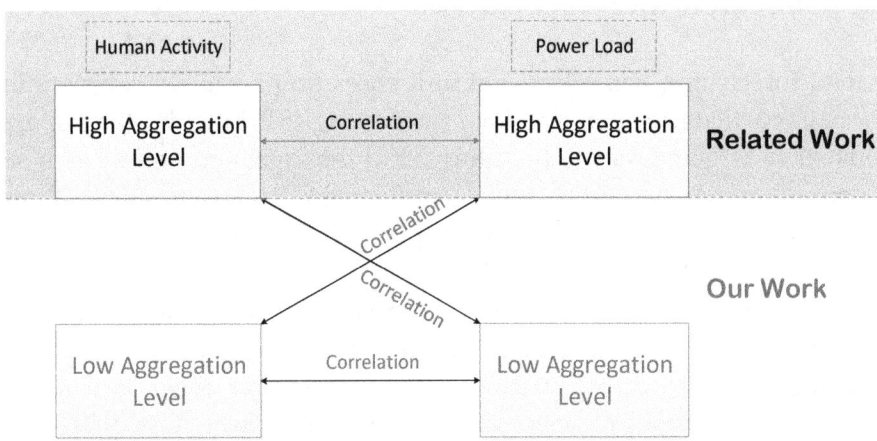

Figure 50.: Comparison of our work with existing research work

6.3 THE FRAMEWORK WORK FLOW

Our STLF framework follows a number of systematic procedures. In general, there are five basic steps as depicted in Figure 51: 1) collecting data, 2) preprocessing data, 3) building the forecasting model, 4) train, and 5) test performance of model. As shown in the figure, a set of learning algorithms are integrated in the framework, in order to compare different forecasting results.

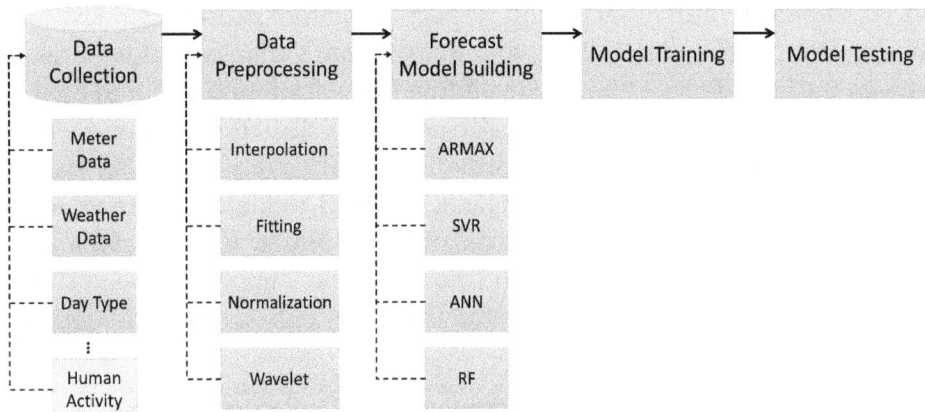

Figure 51.: Basic flow of our load forecasting framework

6.3.1 *Data Collection*

Collecting and preparing input sample data is the first step in designing STLF models regarding the feature vector. Historical measurement data of power consumption is considered as the primary input data: for hourly and daily load forecasting, we use load information of past 24 hours and past 7 days as inputs, respectively. Hourly or daily weather conditions, such as wind speed, cloud cover, temperature and humidity, can be optionally introduced into STLF models depending on the availability of the weather data. Another state-of-the-art influencing factor that we include in the framework, is weekday type. The weekday type input indicates the calendar information (weekdays or weekends).

As we propose an activity-aware STLF framework, the most challenging task of such a framework is to integrate the available or recognized human activity information. In general, the framework should support different activity recognition systems that extract and prepare activity information from for instance a sensor-based recognition system. Later in Section 6.4, we introduce a concrete use case for the framework design in terms of the activity recognition system within a *Smart Office* environment.

6.3.2 *Data Preprocessing*

Following data collection, a data preprocessing step is required to "clean" power load data through: 1) solving the problem of data outliers or missing data, 2) normalizing data and 3) transforming data.

First, data outliers or missing data will be interpolated, e.g. replaced by the average of neighboring values during the same day (for an hourly forecasting) or the same week (for a daily forecasting), or replaced by the value of the same hour and the same weekday of the past week (for an hourly forecasting) or the value of the same weekday of the past week (for a daily forecasting). Since mixing input variables with large magnitudes and small magnitudes will confuse the learning algorithm on the importance of each variable and may force it to finally reject variables with a smaller magnitude [24], the input sample data and the corresponding target vector for both forecasting model training and testing need to be normalized for the feature vector. Finally, the normalized load data is wavelet transformed during the preprocessing step, since wavelets are able to extract redundant information and periodic behavior from load data and improve forecasting accuracy [2].

6.3.3   Building Forecasting Models

After the input sample data in terms of a feature vector is normalized with or without a wavelet transformation, different prediction and regression methods are incorporated in this framework for constructing a STLF. For each forecasting algorithm, a set of necessary configuration parameters can be specified individually and properly by users.

- *A Common Parameter:* a parametrization for the number of previous load data that is taken into account in the feature vector.

- *Autoregressive Moving Average Model with Exogenous Inputs (ARMAX) Model:* a parametrization for the number of autoregressive terms, the number of moving average terms and the number of exogenous inputs terms.

- *Support Vector Regression (SVR):* a parametrization for the cost of error C, the width of the ϵ-insensitive tube, the mapping function ϕ.

- *Artificial Neural Network (ANN):* a parametrization for the number of hidden layers, neurons in each layer, activation function in each layer, training function, number of training epochs, mean squared error goal and spread of radial basis function.

- *Random Forest (RF):* a parametrization for the number of regression trees and the randomly selected feature ratio.

6.3.4   Training and Testing Forecasting Models

The above forecasting models refer to a traditional time-series-like prediction methodology, which means that they are rather data-driven approaches than model-based ones. During the training process, the parameters of each forecasting model will be estimated based on a training dataset of the given time series. In order to avoid an over-fitting, cross-validation is applied. The input sample dataset is split into a training dataset and a validation dataset.

The last step is to test the performance of each trained forecasting model. At this stage, we validate and test the models with the above split validation dataset. The criteria used for defining the measure of forecasting error are the *Mean Absolute Percentage Error (MAPE)* and the *Mean Square Error (MSE)*. MAPE is a standard for examining the quality of load forecasting models, while MSE provides information on the short-term performance as a measure of the variation of predicated values around the measured data. The lower the MAPE and the MSE are achieved, the more accurate is the forecasting. The mathematical expressions of these both measures are given by:

$$\textbf{MAPE} = \frac{1}{N} \sum_{i=1}^{N} \left| \frac{P_a(i) - P_p(i)}{P_a(i)} \right| \times 100 \tag{68}$$

$$\textbf{MSE} = \frac{1}{N} \sum_{i=1}^{N} (P_a(i) - P_p(i))^2 \tag{69}$$

where $P_a(i)$ and $P_p(i)$ denote time series of the actual load and the predicted load, respectively. N is the number of hours or days, which depends on the lead time of the required forecasting.

6.4 THE FRAMEWORK DESIGN

This section provides information about the design phase of the proposed activity-aware load forecasting framework, which follows a top-down approach. It starts with the framework as a whole and then inspects the individual components that comprise the framework. For the concrete use case in this chapter, we introduce our activity recognition system within a smart office environment that meets the following requirements for the framework:

- There should be an office environment equipped with sensors that continuously monitor the environment.

- Hardware and software infrastructure should be available for reading and collecting sensor data. This component is also called as the sensing infrastructure.

- An activity recognition component should process sensor readings to extract activity information.

- A load forecasting component should utilize the recognized activity information.

Based on these requirements, we define the system components as depicted in Figure 52. In this high-level framework architecture, the arrows between the components represent the information flow and are labeled with the type of information that is exchanged between the components:

- The sensing infrastructure collects readings from various sensors in the environment.

- The activity recognition component obtains required sensor data from the sensor infrastructure to recognize activities.

- The load forecasting component obtains required activity information from the activity recognition component to improve the load forecasting process.

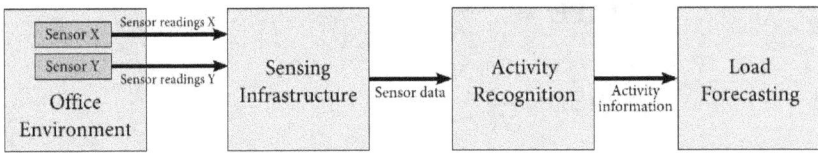

Figure 52.: Initial high-level framework design

6.4.1 *Smart Office Environment*

A smart office environment is defined as a workplace where a set of sensors and actuators are installed for monitoring the environment continuously and controlling the environment based on the sensor data in order to create a better working experience for employees [21]. The infrastructure of our smart office environment consists of heterogeneous wireless sensor networks (WSNs) that include the following three layers:

Sensing controller: The server software that manages sensors and actuators, and collects data from the sensors, and controls the actuators accordingly.

Base station: Sensor/actuator controller hardware that acts as a translator for the communication between heterogeneous sensor networks and the sensing controller. Base stations are device specific, meaning that there can be different types of base stations for different types of sensors.

Sensor nodes: Actual sensor hardware. New types of sensors can be included in the framework system as long as an appropriate base station is available.

6.4.2 *Sensing Infrastructure*

The sensing infrastructure is defined as a component that receives and stores the sensor data from the smart office environment. It acts as a central target for all types of sensors and a central data source for the data processing components that are introduced in the high-level architecture in Figure 52, such as the activity recognition component and the load forecasting component. In order to manage and store sensor data received from different types of sensors in a more flexible and extensible way, we decide to install minimal preprocessing units for every sensor type. These units handle different types of sensor data, which in turn are converted into a uniform data format and sent to the central sensing controller of the infrastructure. Thus, the sensing infrastructure must meet the following requirements:

- The server software should abstract the underlying data storage engine and offer a standardized REST interface for querying and persisting data.

- It should be possible to change the data storage engine; therefore, the data storage engine should not be accessed directly, only using the interface defined in a).

- An easy-to-use web interface should be provided, which is decoupled from the server. The web interface should only use the public REST interface defined in a) and provide real-time monitoring and sensor management.

- The server software and the web interface should be built in a modular manner to allow the framework integrated into them without much effort.

An overview of the architecture is depicted in Figure 53.

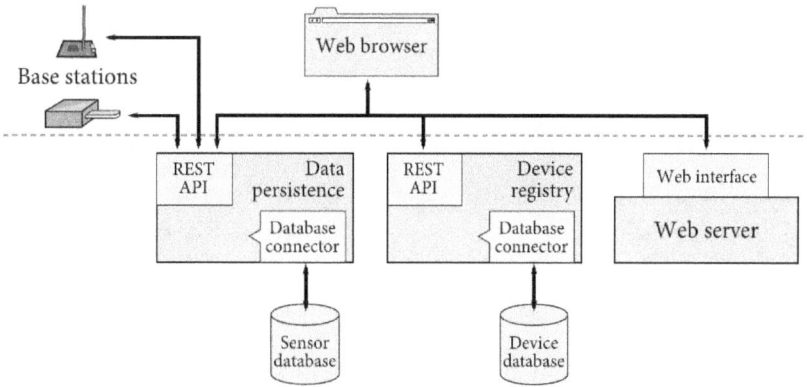

Figure 53.: Architecture design of the sensing infrastructure

The infrastructure consists of two distinct servers, namely a device metadata registry server and a data persistence server. This increases not only the maintainability of these two entities but also the reliability, in a way that a fault in one of the servers cannot disrupt the availability of the other one. The metadata registry and the data persistence servers have their own data storage engines. To achieve the goal of not being dependent on a single data storage solution, the database accesses in these servers are abstracted, so that only the most generic way of accessing and modifying data is through defined interfaces. Database connectors are determined to implement these interfaces by using database solution specific commands and behaviors. Therefore, to support a new database solution, only a new database connector is needed, which implements the associated database interface. Furthermore, these two servers are configurable using configuration files that are outside of the application package. The configuration files offer a possibility to choose which implementation should be used for the modular parts of both servers. For instance, the desired database connector and the RESTful service endpoints can be configured flexibly using the configuration files. The implemented data persistence server transforms the received sensor data in JSON (JavaScript Object Notation) format and uses a data persistence configuration file for the OpenTSDB connector that allows selecting the database engine host of OpenTSDB version 2[1]; while the device registry server manages all sensors and actuators with the internal representation of devices in an entity-relationship model (see Figure 54) and uses a device registry configuration file for selecting the SQLite device database connector to access the SQLite database engine.

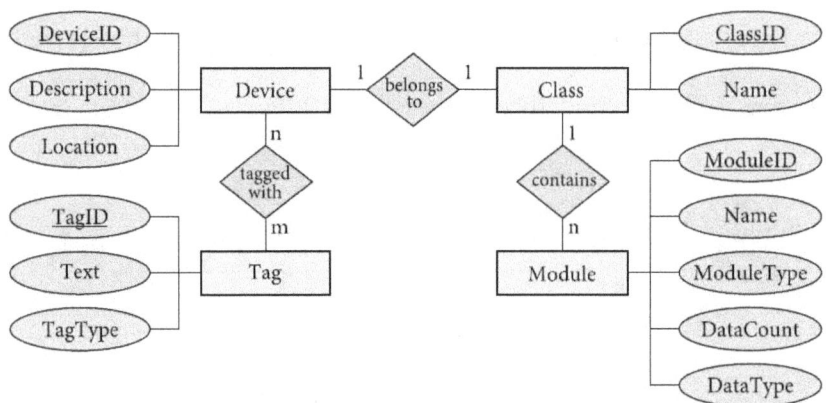

Figure 54.: Device registry entity-relationship model

The web interface is completely separated from the server software and hosted on its own web server. Loosely coupling the web interface makes it possible to develop and update it separately from the server software. The web interface is designed as a single-page web application with an object-oriented user interface using HTML and JavaScript. For every entity in the infrastructure, a model object is implemented in a way that the internal data structures of the models resemble the entities that are defined in the infrastructure, so that actual objects can be automatically fetched and modified using the REST interfaces of the device registry and data persistence servers. Furthermore, various visual representations and interactions are implemented as views, and their appearances are defined and

---

1 An open-source time series database engine offers a powerful REST interface for accessing data: http://opentsdb.net/

designed using the HTML mark-up language and the CSS style-sheet language. Controllers are implemented for managing different parts of the web application that are shown in Figure 55.

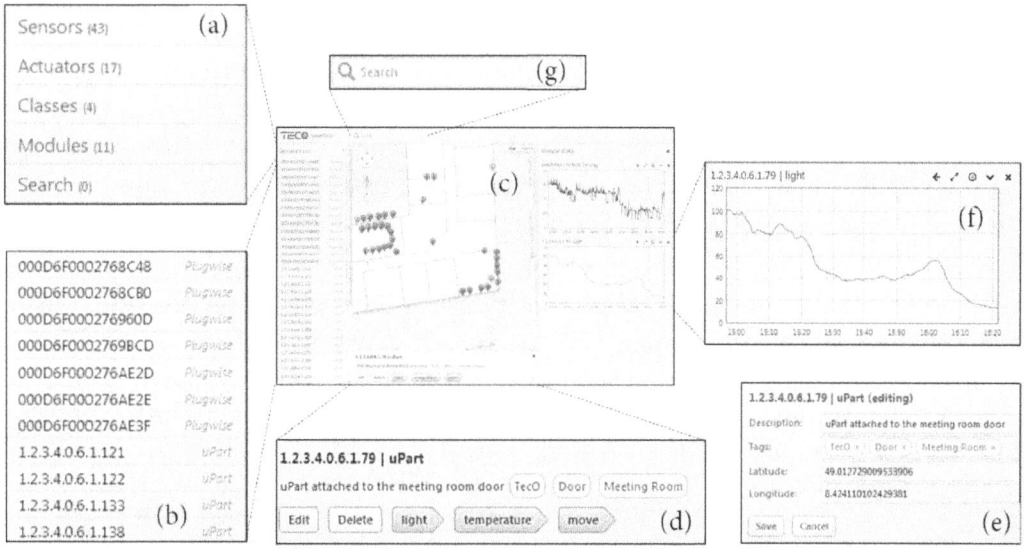

Figure 55.: Management web interface showing deployed sensor on the floor plan of our trial setting

Figure 55 shows different parts of the web interface. On the upper left, a drop-down box (a) can be used to select the types of entities that should be listed in the list below (b). This list is automatically populated with all the objects that belong to the selected entity type. The map (c) in the center is synchronized with this list, in which every object is represented by a marker on the map. Selecting a device from the list or from the map opens a device view (d) on the middle bottom of the figure. This view shows the ID, type, description and tags of the selected device. Using the buttons in the device view (d), the device can be edited or deleted. Clicking the edit button reveals the editing view marked with (e). Furthermore, modules of the selected device are listed as buttons, and clicking a sensor button opens the real-time monitoring view (f) on the right side of the figure. The web interface also includes a search function, with which the given text in device IDs, descriptions and tags can be searched. The search results are then shown in the list and on the map.

6.4.3 Activity Recognition

During a related work research on activity recognition, we noted that many different methods and approaches can be used to recognize human and group activities. Therefore, it is decided that the activity recognition methods should be as customizable as possible in this framework, meaning any activity recognition method or approach should be easily integrated into the module. The requirements for the activity recognition component can be then summarized as follows: The activity recognition module should be flexible as possible and consist of pluggable "sockets" that represent different steps of the activity recognition process, and different "plugs" or interchangeable parts that represent a method, approach or an extra step whose internals can be defined freely. The only hard-coded functionality is the data retrieval task from the data persistence server, and the coordination of various interchangeable parts so that the activity recognition process can run from start to end without a user intervention.

To be able to parametrize and set conditions for the recognition procedure of an activity, the concept of "domain" is introduced. A domain keeps all the needed information for recognizing an activity, such as sensors associated with the activity or the recognition method. A domain is specified by a domain definition file. Figure 56 shows the domain definition file for the domain "meeting" as an example.

```
{
    "name": "meeting",

    "features": [{
        "name": "projector",
        "id": "000D6F0000D34235",
        "module": "energy"
    }, {
        "name": "door",
        "id": "1.2.3.4.0.6.1.79",
        "module": "light"
    }],

    "features-aux": [{
            "name": "daytime", "column": "daytime"
        }, {
            "name": "calendar", "column": "weekday"
        }, {
            "name": "calendar", "column": "is_holiday"
        }, {
            "name": "calendar", "column": "is_workday"
        }
    ],

    "windowsize": 900,

    "featuregenerator": "generators/generator_std.py",

    "modelbuilder": "models/mb_decisiontree.py"
}
```

Figure 56.: Example of a domain definition file

The domain definition file is serialized in a file as a JSON object. The individual definitions in the file are described as follows, which bring the activity recognition process in one picture as shown in Figure 57.

- First, which is, in our case, the name of an energy relevant activity. There can be different procedures and approaches to recognize an activity, so each one of them should be defined by different domain definition files with a unique domain name.

- The features array lists sensors that are associated with the activity. These sensors act as data sources. In the feature generation step, a feature generator of choice will automatically retrieve and process the data from these sensors.

- The auxiliary features array defines the list of external data that should be included in the feature generation.

- The activity recognition process uses a fixed window size, which means that the input is sensor data and external data within the interval of a given time window, and the resulting activity information belongs to the given time window. The window size is specified in seconds.

- The feature generator property is used to select the source file which includes the desired implementation of a feature generation. In our case, the source file is written in Python.

- The model builder property is used to select the source file which includes the implementation for building the recognition model. In our case, the source file is also written in Python.

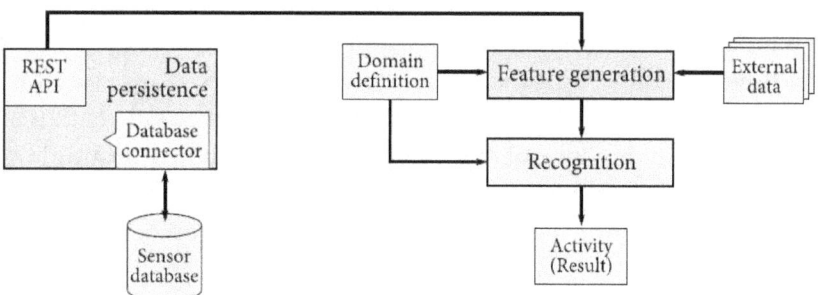

Figure 57.: Activity recognition overview in the proposed framework

The feature generator approach is flexible as well, which means only the parts that must not be changed are located in a core feature generator program. For

171

other tasks, the feature generator core offers the generator function a number of other functions along with the parsed domain information, which can be accessed by the generator function flexibly. These functions and objects are listed in Table 9. The domain object includes the same information from the domain definition file, and is a Python dictionary. Sensors and external data sources are specified in form of lists in the domain object, so that the generator function can iterate the lists and create features. Moreover, the generator function runs in a sandbox, which protects the internals of the feature generator core from an access by generator functions, and adds the aforementioned objects and functions to the context of the generator function. An example generator function can be seen in Listing 6.1.

Table 9.: Functions and objects that are available to the generator function

Name	Type	Description
`domain`	Object	Domain definition
`fetch_data(sensor_id, module)`	Function	Retrieve sensor data
`fetch_aux(source, data_name)`	Function	Retrieve external data
`add_feature(feature_name, value)`	Function	Add a feature

For every entry in the list of `features` in the `domain` object, the `generate-_standard` function in Listing 6.1 generates the features for a single sensor input. Also other functions are allowed to be defined internally in the generator function for generating different features for different sensors. Given a sensor data source, the `generate_standard` function retrieves the data using the `fetch_data` function. Meanwhile, the feature generator core takes over and retrieves the sensor data for the current time window by accessing the data persistence server, afterwards interpolates it and returns the data to the generator function. The generator function is time window invariant, which means that it does not know for which time window features should be generated. The feature generator core manages the time windows and ensures the generator function with the relevant data of the correct time window. After the sensor data is retrieved, the generator function applies a number of functions over the data to create feature values that are added to the feature list for the current time window. In the above example generator function, four features are generated for every sensor data source: *arithmetic mean, standard deviation, arithmetic mean of the Fourier coefficients* and *standard deviation of the Fourier coefficients*.

Moreover, the purpose of the model builder function is to flexibly create and return any machine learning model in terms of a Python object that consists of two functions, i.e. `train` and `predict`. An example model builder function can be seen in Listing 6.2. Besides the historic feature vector instances, the `train` function of the model requires a second parameter that refers to `annotations`.

In our case, an annotation is the activity information associated with a feature vector. Based on the feature vector instances and the annotations, the model is then trained with either a machine learning library or self-implemented functionality. In the example model builder function, a decision tree classifier from the scikit-learn library is applied.

Listing 6.1: A generator function example

```python
import numpy as np

def generate_standard(feature_name, sensor_id, module):
  data = fetch_data(sensor_id, module)
  if len(data[1]) == 0:
    add_feature(feature_name + "-mean", 0)
    add_feature(feature_name + "-std", 0)
    add_feature(feature_name + "-fft-mean", 0)
    add_feature(feature_name + "-fft-std", 0)
  else:
    add_feature(feature_name + "-mean", np.mean(data[1]))
    add_feature(feature_name + "-std", np.std(data[1]))
    data_fft = np.fft.fft(data[1])
    add_feature(feature_name + "-fft-mean",
      np.mean(abs(data_fft)))
    add_feature(feature_name + "-fft-std",
      np.std(abs(data_fft)))
for feature in domain["features"]:
  if feature.has_key("generator"):
    generator = feature["generator"]
  else:
    generator = "generate_standard"
  locals()[generator](domain["name"], feature["name"],
    feature["id"], feature["module"])
if "features-aux" in domain:
  for aux in domain["features-aux"]:
    v = fetch_aux(aux["name"], aux["column"])
    add_feature(domain["name"], aux["name"] + "_"
      + aux["column"], v)
```

Listing 6.2: A model builder function example

```python
from sklearn import tree

class DecisionTreeModelBuilder(object):

  def __init__(self):
    self.model = tree.DecisionTreeClassifier()

  def train(self, feature_vectors, annotations):
    self.model.fit(feature_vectors, annotations)

  def predict(self, feature_vector):
    return self.model.predict(feature_vector)

model = DecisionTreeModelBuilder()
return_model(model)
```

6.4.4 *Load Forecasting*

The load forecasting process is also specified and parametrized in a domain definition file, which means that for every appliance, a domain should be set up. It is also possible to use the same domain for activity recognition as well as for load forecasting, meaning that using the same domain definition, the activities in this domain can be recognized and the future load of appliances in this domain can be predicted. For this purpose, extra parameters and definitions are appended to the domain definition file, and an example can be seen in Figure 58.

The feature generator from the activity recognition component is employed and appropriately modified to additionally produce a feature vector based on these parameters for the training phase of the load forecasting process. The output of the feature generator as a feature vector consists of the mean load value from the sensor specified with "lf-features" in the time window t, the external factors specified with "lf-features-aux" in the time window t and the mean load value of the m previous time windows from the same sensor specified with "lf-previousload".

As mentioned before, we aim at a load forecasting process by appending the recognized activity information into the feature vector, which means that the energy-relevant activity that is supposed to improve the forecasting accuracy should be firstly identified. For this purpose, the framework offers the functionality to analyze correlations between human activities and mean load values from different

174

domains. The domain for which the load should be predicted is chosen as the source of the load data. Then, the load data from this domain is analyzed against activities from other domains. For an activity-load pair, the correlation analysis with different approaches, e.g. Pearson correlation coefficient, can be conducted multiple times by shifting the activity time window as well as by increasing the number of time windows between the activity and the load. For instance, the load in the time window t is tested for correlation with the activity in the time window $t-1$, $t-2$, $t-3$ and $t-4$. This process is repeated for the complete data, and the results are evaluated. If a high correlation is found between the load and an activity from another domain with a specific time shift, this activity will be included in the feature vector used in the load forecasting process. In comparison to the classic load forecasting, the feature vector for predicting the load at the time window t is then modified as shown in the following expression.

$$
F_t = \begin{pmatrix} EF_1(t) \\ \vdots \\ EF_n(t) \\ P(t-1) \\ \vdots \\ P(t-m) \end{pmatrix} \rightarrow F_t = \begin{pmatrix} A \\ EF_1(t) \\ \vdots \\ EF_n(t) \\ P(t-1) \\ \vdots \\ P(t-m) \end{pmatrix} \tag{70}
$$

where $EF_x(t)$ are the external factors and $P(t-y)$ are the mean power load value of the previous time windows, while A is the relevant activity information.

6.5 STLF EVALUATION WITH ACTIVITY RECOGNITION

In this section, we conduct an hour-ahead STLF analysis based on the above framework for the power consumption of a projector in the meeting room in our smart office environment, which refers to the domain "meeting". In this case study, we consider relevant activity information from the activity recognition component. The power consumption of a projector refers to a pure appliance-level load, which is visualized in Figure 59.

First, based on the classic load forecasting approach, in which only historic load information and external factors (i.e. daytime and day type) are considered, the following two load values from previous time windows are determined to produce the accurate result for this case study when the load for the time window t needs to be predicted:

- Load from time window $t-1$ (Load from last hour)

```
"lf-features": [{
    "name": "projector",
    "id": "000D6F0000D34235",
    "module": "energy"
}],

"lf-features-aux": [{
        "name": "daytime", "column": "hour"
    }, {
        "name": "daytime", "column": "daytime"
    }, {
        "name": "calendar", "column": "weekday"
    }, {
        "name": "calendar", "column": "is_holiday"
    }, {
        "name": "calendar", "column": "is_workday"
    }
],

"lf-featuregenerator": "generators/lf_generator_std.py",

"lf-windowsize": 3600,

"lf-previousload": 2
```

Figure 58.: Load forecasting parameters in a domain definition file

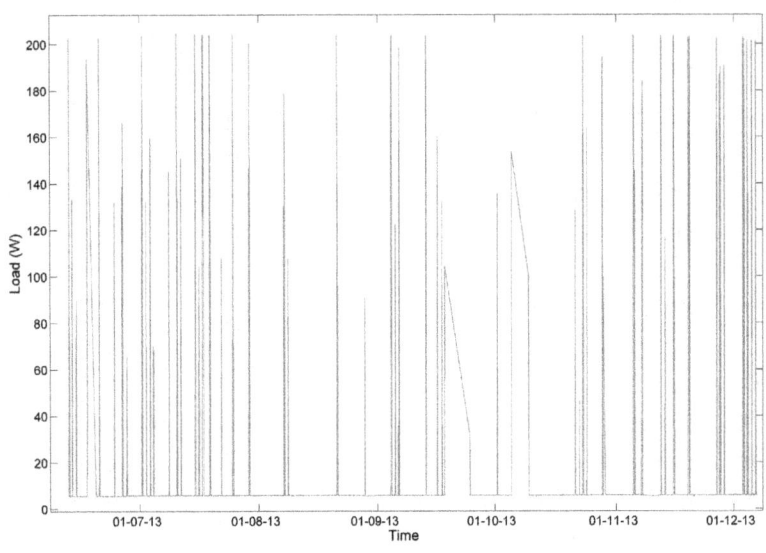

Figure 59.: A example load curve of the correspondent projector

- Load from time window $t - 2$ (Load from two hours before)

Then, the 2/3 of the time windows are used for training classic load forecasting models without any activity information. The remaining 1/3 are predicted on an hourly basis and compared with the actual load values. The forecasting results of this classic load forecasting approach for the domain "meeting" are shown in Table 10.

Table 10.: Comparison of forecasting results for the domain "meeting" in terms of MAPE and MSE

Algorithm	MAPE (%)	MSE
Multiple linear regression	63.712	754.494
Regression tree	39.088	836.786
Support vector regression	17.764	891.126
Artificial neural network	41.671	719.238
Kalman filter	65.050	847.155

As seen in the table, different forecasting algorithms reveal different results regarding the forecasting accuracy and no forecasting algorithm provides an accurate appliance-level load forecasting in terms of both MAPE and MSE. Due to infrastructure downtimes during the data collection, the quality and quantity of the load data are negatively affected. That's why that only SVR delivers a reasonable forecasting accuracy in terms of MAPE (17.764%), as SVR exhibits its good learning ability for a small set of training data.

In the next step, we integrate relevant activity information to the load forecasting process. For this purpose, a correlation analysis between activities and the load of the projector is conducted. The correlation result shows that only one of the activities, i.e. the "drink" activity is slightly correlated with the load of the projector. The correlation coefficients between the load and other activities are close to zero. The correlation coefficients between the load and the "drink" activity are depicted in Figure 60. The "drink" activity is defined as the *number of people who currently consume or just consumed a hot drink*, which means the value of the "drink" activity is accumulated to the next time windows. In our case study, for calculating the activity value, next 3 time windows are taken into account. As shown in the figure, the greatest correlation coefficient is at the time window difference of 4 windows, i.e. 1 hour difference. That means that the "drink" activity has the strongest relationship with the load of the projector after one hour.

Before we demonstrate the load forecasting results, an overview of the activity recognition result for the domain "drink" is presented in Table 11. The domain "drink" for the activity recognition component is specified as the *number of hot*

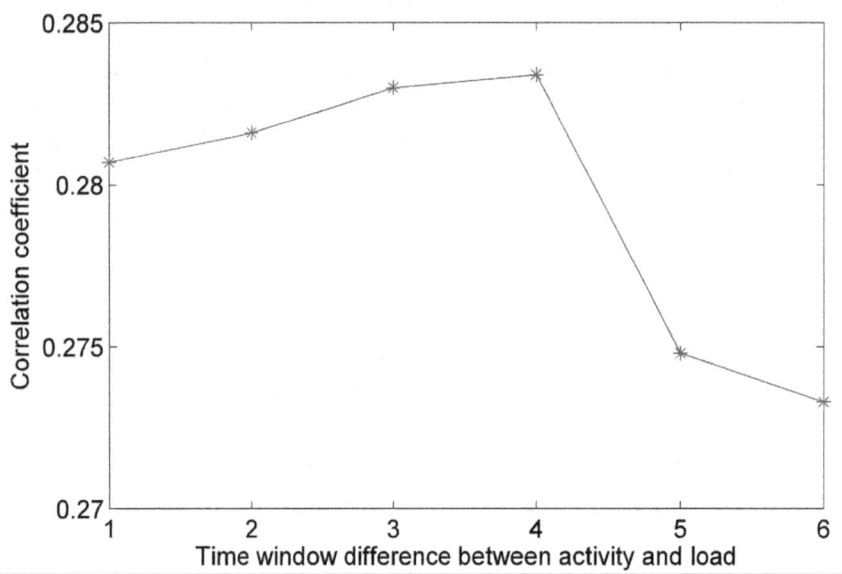

Figure 60.: Correlation coefficients between the "drink" activity and the projector load

drinks prepared in a time window including coffee and hot water. In total, 7 activity classes are defined with 0 meaning no drinks prepared in the time window, 1 meaning one drink prepared, and so on. The confusion matrix in Table 11 reveals in general a good accuracy of the activity recognition for the domain "drink".

Subsequently, based on the recognized "drink" activity values, the load information of time window $t-1$ and $t-2$ as well as the correspondent external influencing factors, different load forecasting models are trained with the 2/3 of the feature data. The remaining 1/3 of the data is predicted and compared with the actual load values. The forecasting results of this activity-aware load forecasting approach for the domain "meeting" are summarized in Table 12 and visualized in Figure 61.

In comparison to the forecasting results without the activity information in Table 10, we observed that including activity information does not improve the load forecasting accuracy for our case study. However, the reason can be 4-fold:

- A correlation coefficient of 0.28 can be considered as a weak correlation, which means that in our case, there is no any strong relationship between the load and the activity that can improve the load forecasting.

- Using the available sensors, only a small number of activities can be defined, which leads to a small number of possible correlations. Among the activities

Table 11.: Confusion matrix of the recognition results with a random forecast classifier for the domain "drink"

Actual \ Predicted	0	1	2	3	4	5	6
0	6009	2	0	0	0	0	0
1	13	506	25	0	0	0	0
2	0	36	152	12	0	1	0
3	0	1	18	35	1	1	0
4	0	0	1	8	4	2	0
5	0	0	0	0	1	0	0
6	0	0	0	0	1	0	0

Table 12.: Comparison of forecasting results for the domain "meeting" in terms of MAPE and MSE

Algorithm	MAPE (%)	MSE
Multiple linear regression	65.105	748.234
Regression tree	41.072	836.767
Support vector regression	17.370	937.472
Artificial neural network	41.666	883.956
Kalman filter	68.615	845.627

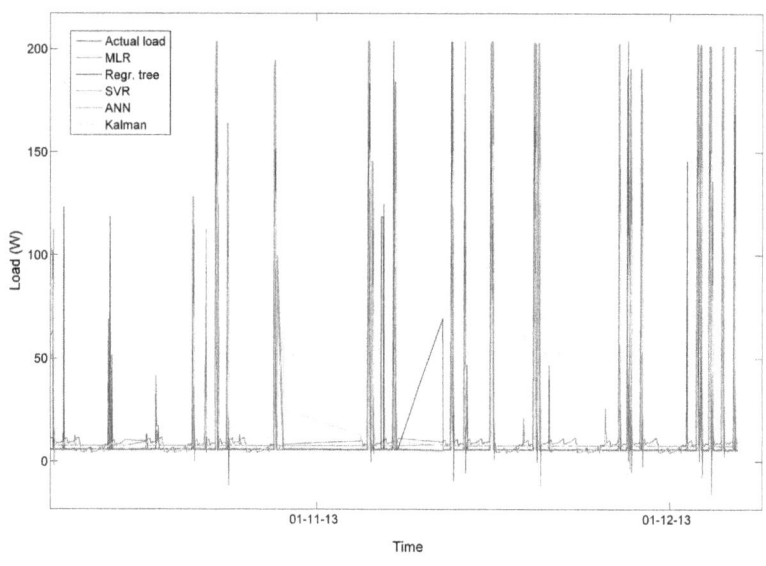

Figure 61.: Visual comparison of forecasting results for the domain "meeting"

defined for the case study, only one activity is in a weak correlation to the load of the projector.

- The quality and the quantity of the collected data are negatively affected by infrastructure downtimes.

- A load forecasting at appliance level is in any case difficult due to the strong irregularity of the appliance-level power consumption.

Although no expected improvements in the load forecasting accuracy can be observed, the framework still proves its worth for analyzing the correlation between activities and loads as well as integrating the activity information into the load forecasting process.

6.6 STLF EVALUATION WITHOUT ACTIVITY RECOGNITION

In this section, two further forecasting studies based on the above framework are conducted. Both studies focus on load forecasting at meter level, and utilize existing or prepared activity information at different scales instead of recognizing activities.

6.6.1 *Case Study 1*

The first case study refers to the database of smart meter measurements taken in the NOBEL[2] project. The database includes 15-minute interval measurements for about 5000 meters taken between November 2010 and February 2013 in the project's field trial in Alginet, Spain. The raw dataset contains outliers, as well as missing meter readings due to downtimes or other infrastructural issues, see Figure 62. For the STLF task, data of about 2/5 meters, i.e. about 2000 meters, can be used. Furthermore, due to European Data Protection Regulation and privacy concerns, the NOBEL dataset can not provide our desired level of quality in terms of forecasting service. In addition, the anonymization process performed on this dataset has further eliminated some useful features for prediction purposes, including *location, consumer type*, etc. Based on the aforementioned issues, it seems reasonable to filter the dataset and also select those meters which contain "sufficient" readings, i.e. consumption data ranging from November 2010 to the beginning of February 2013. Therefore, it is determined to consider only 10 smart meters to evaluate our STLF framework.

The meter readings are interpolated into hourly and daily power consumption values and subsequently wavelet transformed. Then, we employ the *ARMAX*,

---

2 http://www.ict-nobel.eu

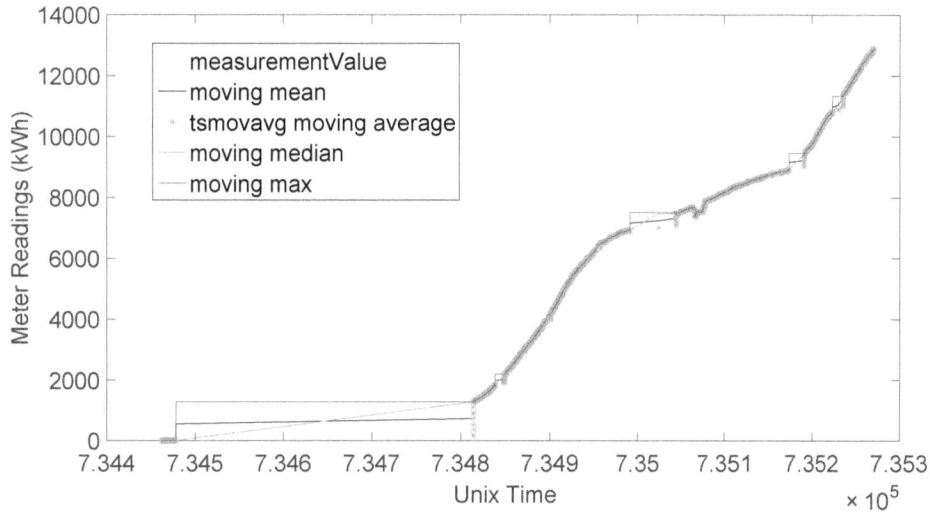

Figure 62.: The raw dataset of a smart meter as well as the moving average of the raw data

SVR, *ANN* and *RF* respectively combining them with independent influencing factors, such as different weather conditions (incl. *temperature*, *humidity*, *pressure* and *wind speed*), weekday types (weekdays and weekends) and big events. For this case study, we cannot collect any human activity data from the proposed activity recognition component. Therefore, instead of fine granular human activity information, we consider big events (e.g. sport events, concerts, etc.) as an additional influencing factor, which are more related to aperiodic human activity than the weekday types and holidays, and can be derived from special calenders of Alginet as well as social media. Finally, we train the load forecasting models with a two years dataset and test the models with the rest of the data.

We separate the evaluation of the forecasting results through the following combination cases of influencing factors: 1) without any influencing factor; 2) weekday types only; 3) temperature only; 4) humidity only; 5) pressure only; 6) wind speed only; 7) big events only.

For the above 7 cases, we evaluate not only individually the selected 10 smart meters but also the aggregation of 10 smart meters. The prediction metrics in Table 13 indicate the average MAPE and MSE of the selected 10 meters, while Table 14 shows the MAPE and MSE of the aggregated power consumption prediction. Since we reconstruct the consumption data from the normalized and wavelet transformed one, which ranges in the real life from about $40Wh$ to about $650Wh$, see e.g. Figure 63 and 64. This is the reason for that the MSE of each approach reaches at least 10^2 (one meter scenario) or 10^3 (aggregation scenario) for 36 days ahead forecasting. We notice that in general the forecasting results at the aggregation level are more accurate than the ones for an individual smart meter. Moreover, the MAPE values in Table 13 show that the influencing factors except *wind speed* (case 6) and *big events* (case 7) can lightly improve the prediction accuracy for the single smart meter scenario. However, the influencing factor *big events* contributes to the accuracy improvement for the aggregation scenario, see case 7 in Table 14. Big events, which correspond to human activity at the higher aggregation level, show an impact on the load forecasting of aggregated metering data. In order to prove our hypothesis that the aggregation level of meter data and activity data matters, the next case study focuses then on the human activity data at household level to show its impact on the load forecasting for one individual smart meter.

Model	ARMAX	SVR	ANN	RF
1) MAPE	20.98	22.53	22.58	19.45
MSE	175.87	213.28	210.23	166.54
2) MAPE	20.78	22.38	25.17	18.85
MSE	172.05	212.04	252.21	156.40
3) MAPE	21.26	21.90	22.88	18.44
MSE	180.17	198.55	211.71	146.75
4) MAPE	20.88	22.49	22.70	19.54
MSE	173.22	212.71	246.74	163.75
5) MAPE	20.76	22.54	25.46	19.35
MSE	173.42	213.60	322.21	160.99
6) MAPE	21.33	22.94	24.65	19.78
MSE	179.60	219.58	253.03	169.61
7) MAPE	21.08	22.58	26.27	19.55
MSE	177.63	213.93	285.86	163.91

Table 13.: Average MAPE (%) and MSE of 10 smart meters for different forecasting models in different cases

Moreover, Figures 63 and 64 depict the prediction results for one representative smart meter of the 10 selected meters and the aggregation of 10 metering data, regarding case 7) with the influencing factor *big events*, respectively. From the predicted load curves in both scenarios, we notice that *RF* and *ARMAX* approaches both delivered better load forecasts in comparison to the other approaches. By means of the only influencing factor *big events*, the comparison of both figures shows again that high-level human activity information influences the power consumption at higher aggregation level rather than at individual meter level. However, overall, the improvement of the forecasting accuracy by additional information of any above defined influencing factors is not significant.

Model	ARMAX	SVR	ANN	RF
1) MAPE	12.32	12.94	11.15	10.51
MSE	$5.61 \cdot 10^3$	$6.60 \cdot 10^3$	$4.74 \cdot 10^3$	$4.42 \cdot 10^3$
2) MAPE	12.69	13.06	14.65	10.04
MSE	$5.44 \cdot 10^3$	$6.57 \cdot 10^3$	$7.72 \cdot 10^3$	$4.57 \cdot 10^3$
3) MAPE	12.67	12.90	12.39	9.57
MSE	$5.54 \cdot 10^3$	$6.54 \cdot 10^3$	$5.62 \cdot 10^3$	$4.25 \cdot 10^3$
4) MAPE	12.25	12.94	18.93	11.51
MSE	$5.36 \cdot 10^3$	$6.61 \cdot 10^3$	$2.32 \cdot 10^4$	$5.81 \cdot 10^3$
5) MAPE	12.76	12.94	57.43	11.04
MSE	$5.80 \cdot 10^3$	$6.60 \cdot 10^3$	$1.09 \cdot 10^5$	$4.71 \cdot 10^3$
6) MAPE	12.39	12.94	13.96	10.75
MSE	$5.40 \cdot 10^3$	$6.59 \cdot 10^3$	$7.68 \cdot 10^3$	$4.50 \cdot 10^3$
7) MAPE	12.20	12.51	10.92	10.37
MSE	$5.26 \cdot 10^3$	$6.51 \cdot 10^3$	$4.13 \cdot 10^3$	$4.39 \cdot 10^3$

Table 14.: MAPE (%) and MSE of aggregated metering data for different forecasting models in different cases

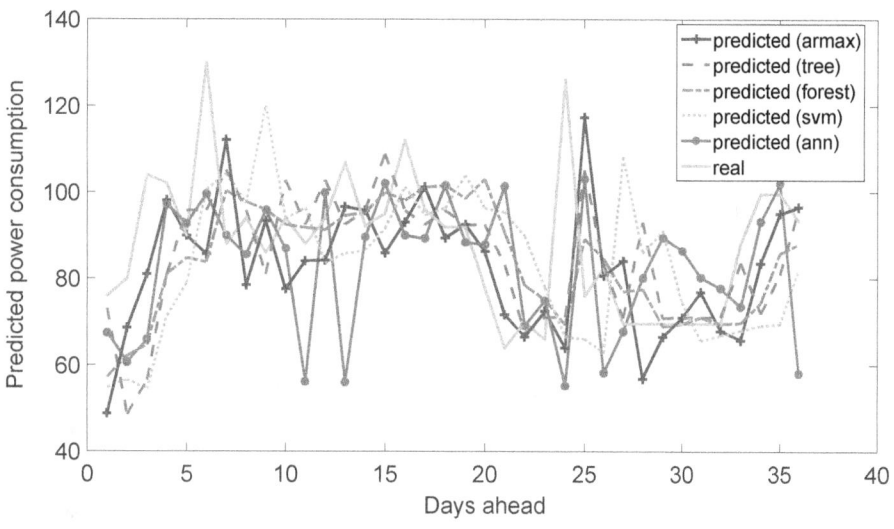

Figure 63.: Comparison of the load forecasting results with influencing factor *big events* for one smart meter

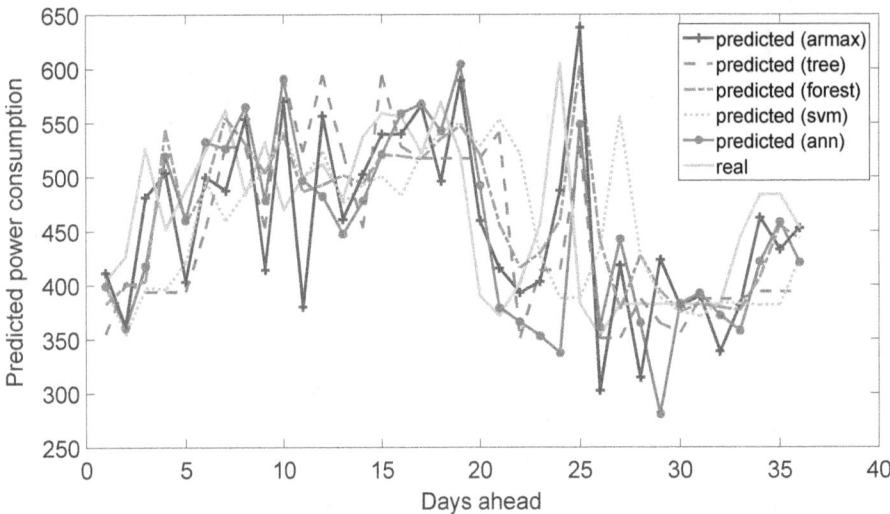

Figure 64.: Comparison of the load forecasting results with influencing factor *big events* for aggregated metering data

6.6.2 *Case Study 2*

The second case study refers to a time-of-use (TOU) dataset of power consumption and activities. The activity information is presented in terms of daily actions of residents. A web-based data logging system records, estimates and edits activity logs in a digital diary style for households. Common types of activities are predefined, but each household can define new activity types that can be reused by other households. Compared to existing approaches of TOU data surveys, the activity types are personal and extensible.

We conducted an experiment over a three-month period from the beginning of December 2014 to the beginning of March 2015. The subjects are voluntary households from diverse places of Japan. They were asked to take a few minutes each day to input their daily activities. As a result, from 21 households, we are able to obtain 10,250 activity data and about 2,369,000 power consumption data in minutes by March 04, 2015. The number of activity types is 52, which comprise indiscriminative activities such as "lunch" and "meal", and also very personal activities. In order to avoid ambiguity for the proposed analysis, we cluster them manually to 11 meta activities: $\mathcal{A}$ = {"at home", "bath", "computer", "cooking", "eat", "hobby", "housework", "media", "out", "relax", "sleep"}.

Correlation Analysis

After reviewing load curves of individual households of our experiment, we could not readily identify any periodic pattern, see example load curves in Figure 65. In order to better understand their load patterns, we first analyze the causal influence of the collected human activities on power load at household level.

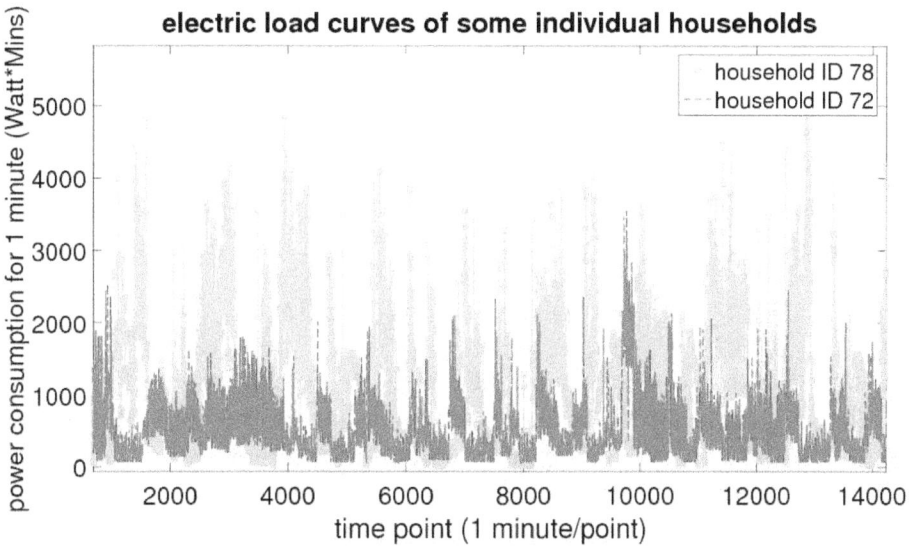

Figure 65.: Two representative households' power consumption measured within the experiment duration

In the following correlation analysis, we investigate the relationship between the power load random variable P and the activity random variable A using a probability- and information-theoretic significance metric. The classical correlation coefficient measures normally the linear relationship (Pearson's correlation) or monotonic relationship (Spearman's correlation) between two random variables. In order to discover non-linearity and uncertainty within the relationship between random variables P and A, we apply for this correlation analysis the mutual information measure that can describe non-monotonic and more complicated relationships between P and A through a weighted sum of their joint probabilities [4]. In comparison, we take additionally the state-of-the-art exogenous influencing factors like day type random variable D and day hour random variable H into account [17]. For conducting mutual information comparison, we introduce some notions of information theory as follows:

- random variable D indicates the day type information that can be one of the 7 categorical values $\in \mathcal{D} = \{$"Sun", "Mon", "Tue", "Wed", "Thu", "Fri", "Sat"$\}$;

- random variable H indicates the day hour information that can be one of the 24 integer values $\in \mathcal{H} = \{0, 1, \ldots, 23\}$;

- random variable A indicates the activity information that can be one of the 11 categorical values $\in \mathcal{A}$;

- random variable P indicates the power consumption information that is numerical $\in \mathcal{P}$, which ranges continuously from P_{min} to P_{max}

The notations of the above random variables are applicable to all households. For each household, we compare the mutual information values between $I(P; A)$ and $I(P; D, H)$, which can be defined as the difference between the entropy $H(P)$ and the conditional entropy $H(P|\bullet)$ as follows:

$$I(P; A) = H(P) - H(P|A), \tag{71}$$
$$I(P; D, H) = H(P) - H(P|D, H). \tag{72}$$

As we know, entropy is an uncertainty measure, which means that the above two equations can be regarded as the reduction of uncertainty in P after observing A or $\{D, H\}$, respectively. However, the mutual information expressions indicate rather the existing of an uncertainty reduction than the significance of an uncertainty reduction. Thus, we utilize them only as the measure of a contribution indication of A or $\{D, H\}$ to estimate P. Since our hypothesis is that the activity information A contributes to the load forecasting, it is going to be approved if

$$H(P) - H(P|A) \gg H(P) - H(P|D, H), \tag{73}$$

which is simplified as comparison of conditional entropy as follows:

$$H(P|D, H) \gg H(P|A). \tag{74}$$

For Inequality 74, $H(P|\bullet) = - \int_P pdf(p|\bullet) \log pdf(p|\bullet) dp$ can be used to calculate both left and right part, which is measured in hartleys due to the logarithmic base 10, where $pdf(\bullet)$ stands for a probability density function. As D, H and A are discrete random variables with given value domain $\mathcal{D}$, $\mathcal{H}$ and $\mathcal{A}$ as described above, respectively, the continuous random variable P will be discretized through quantization in form of data binning with 1 *Watt*. Then the Inequality 74 can be reformed as

$$- \sum_{d \in \mathcal{D}, h \in \mathcal{H}} pmf(d, h) \sum_{p \in \mathcal{P}} pmf(p|d, h) \log pmf(p|d, h) \gg \tag{75}$$
$$- \sum_{a \in \mathcal{A}} pmf(a) \sum_{p \in \mathcal{P}} pmf(p|a) \log pmf(p|a).$$

186

In order to determine the above inequality, we need to estimate two a priori probability mass functions $pmf(d,h)$ and $pmf(a)$, as well as two conditional probability mass functions $pmf(p|d,h)$ and $pmf(p|a)$, for each value combination of day type D and day hour H, and each value of activity type A. Using histogram approach, we count up and normalize the sample numbers of correspondent variables as an approximation of the probability mass functions.

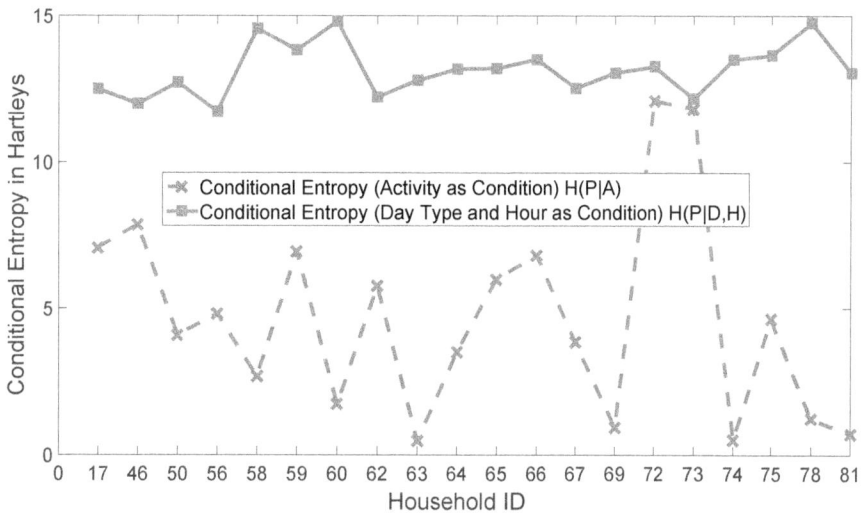

Figure 66.: Results of mutual information analysis

In Figure 66, the conditional entropy of power variable under given day type and hour condition $H(P|D,H)$ and under given activity condition $H(P|A)$, are plotted in gray square and in red dashed, respectively. The household 57 was eliminated for this analysis due to improper activity measurements. Moreover, for all households the conditional entropy $H(P|A)$ is smaller than the conditional entropy $H(P|D,H)$, except the household with ID 73. The reason is that more than 80% of its activity records are dominated by "at home", which highly affects the conditional entropy of other activity types. Therefore, the outcome of this is $H(P|D,H) \gg H(P|A)$ fulfilled for most of the households, thereby, the hypothesis in Inequality 74 is proven.

Forecasting Results

According to the above calculated mutual information measure, the activity information can contribute to the power consumption estimation for certain households, though to varying degrees. In the following, we demonstrate an activity-enhanced load forecasting model with different prediction algorithms, in which the future information of activities is needed.

The collected activity samples (log entries) have very different activity durations in minutes: 99.6% of the sample durations are greater than 10 minutes and 97.1% are greater than 15 minutes. Considering 15-minute-interval meter reading data that are relevant for the energy markets and should be supported by the most of Advanced Metering Systems, our load forecasting model focuses on a prediction lead time of 15 minutes and can be sufficiently evaluated with 97.1% of the activity samples. Furthermore, forecasting experiments with prediction lead time of 30 minutes and 1 hour are conducted to compare the temporal effect of prediction power.

The load forecasting model is trained individually based on the STLF framework using **ARMAX** (Autoregressive Moving Average Model with Exogenous Inputs), **SVR** (Support Vector Regression) and **ANN** (Artificial Neural Network). For each algorithm, we take the discussed state-of-the-art influencing factors day type D and day hour H, and activity information A as input variables into account. The forecasting accuracy is evaluated for the following 3 combination cases of input variables: 1) only D and H as input variables; 2) only A as input variable; 3) all D, H and A as input variables.

The feature vector for training the forecasting model comprises not only the above input values, but also the historical consumption data. In our case, we take the past 24 hours' consumption data for each feature vector into account, which means 96, 48 or 24 past power consumption data points in a feature vector for the load forecasting with lead time of 15 minutes, 30 minutes or 1 hour, respectively. Sample data of 2 weeks for applicable variables are used within each training process. Min-max normalization is applied to all input variables. Day type, day hour and type of activity are unary encoded into the feature vector.

A comparison of the forecasting accuracy is carried out among three proposed prediction algorithms with various settings. We use Mean Absolute Percentage Error (MAPE) as the performance criterion. Firstly, we test ARMAX, SVR and ANN for all individual households with the combination of all applicable activity types by each forecasting run.

Table 15 shows the forecasting results of ARMAX, SVR and ANN with different settings in terms of mean and standard deviation of MAPE values. In this test, we used the first 2 weeks of the dataset to train all three models, and to predict the power consumption of the rest experiment time. As shown in the table, ARMAX could not perform well due to a small quantity of training data. Within ARMAX modeling, we have also taken variable time-delayed inputs into account, on which the current output depends, but no improvement could be seen. In comparison, SVR and ANN exhibit their good learning ability for small samples. In particular,

Models Settings	ARMAX MAPE		SVR MAPE		ANN MAPE	
	mean	std	mean	std	mean	std
15m with D, H	72.2	34.6	44.9	19.0	56.0	27.5
15m with A	123.7	94.6	42.1	17.3	71.0	33.4
15m with D, H, A	78.5	36.9	46.4	20.9	58.5	27.5
30m with D, H	97.4	145.9	50.0	20.5	53.5	24.9
30m with A	$9.0 \cdot 10^3$	$3.8 \cdot 10^4$	50.1	22.1	70.6	33.1
30m with D, H, A	$3.5 \cdot 10^2$	$1.2 \cdot 10^3$	50.3	22.5	62.5	32.2
60m with D, H	70.7	51.3	70.9	73.5	31.2	16.0
60m with A	$7.6 \cdot 10^{10}$	$3.3 \cdot 10^{11}$	73.9	85.2	41.2	24.2
60m with D, H, A	80.7	62.8	69.9	74.3	31.5	17.4

Table 15.: Mean and Standard Deviation (std) of MAPE values (in %) of 21 house-holds for different forecasting models with different settings; D, H and A stand for day type, day hour and activity information, respectively

SVR provides the best results for both 15 minutes ahead and 30 minutes ahead load forecasting. Comparing the mean MAPE values for all the 15m settings as highlighted in green, the SVR forecasting result with activity information as an input variable outperforms all the other. Some issues regarding the authenticity of collected activities may lead to an inconsistency regarding information in case that all D, H and A are considered as input variables for the load forecasting. This is the reason for that in Table 15 the MAPE value in setting with D, H, A is even greater than the MAPE value in setting with D, H. The contribution of activity information for the load forecasting has not only been demonstrated through the mean MAPE value, but also been confirmed in comparison with individual MAPE values as shown in Figure 67. In this bar chart, we notice that for all depicted individual households except for ID 73, the red bar is under the other two bars. That means that considering only activity information as input variable, almost all households can reduce the forecasting error.

For training the SVR forecasting model, we have also tested diverse 2-week data as the training dataset. The results are consistent. Even if we train the SVR fore-casting model with 1 month of data, the results do not show any benefit with respect to forecasting accuracy. Therefore, we focus mainly on the SVR forecast-ing model with 2 weeks training data for the following analysis. As seen in Table 15 and Figure 67, the combination of all activity types as one input variable can improve appropriately the forecasting accuracy for individual households. **What about the impact of each activity type?** Since each household has certain appli-

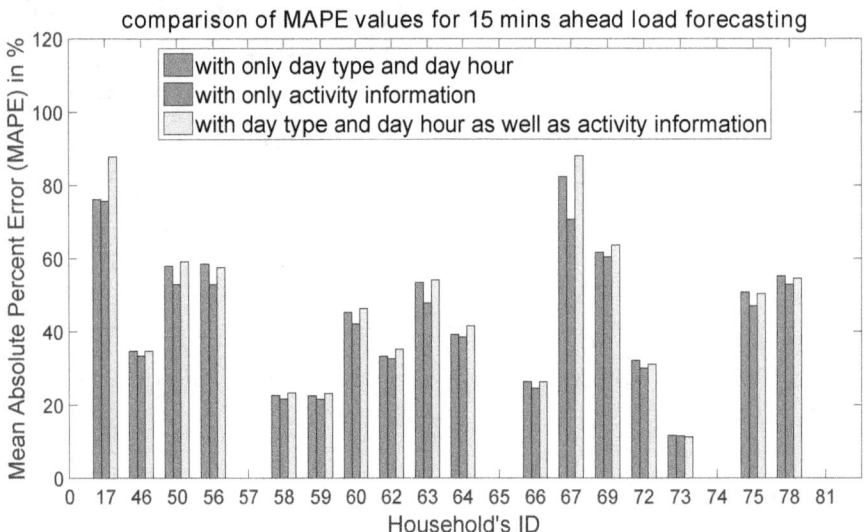

Figure 67.: MAPE comparison of 15 minutes ahead load forecasting with SVR for individual households

cable activity types, Figure 68 shows the impact of 4 individual activity types on the forecasting accuracy of certain correspondent households.

In Figure 68a, we notice that the load forecasting driven by the activity "at home" performs at least the same as driven by the combination of all applicable activities except for the household with ID 46. In comparison, the load forecasting driven by only the activity "cooking" in Figure 68b works well only for the household with ID 17 and 69. Both activity "out" and "sleep" are one of the most conducted activities by almost all households. By comparing the figures 68c and 68d, we notice a common quality for the household with ID 58, 66, 72, 73, 75 and 78 that their red bar is all slightly higher than the green one, which implies that the activity "out" and "sleep" both contribute less than some other applicable activities for these 6 households. To sum up, we found out that the short-term load forecasting (e.g. 15 minutes ahead) at household's level can benefit from individual activities very differently. In other words, the contribution of individual activities for forecasting accuracy is household dependent and household specific.

6.7 AN INTEGRATION TEST

In this section, we utilize again the IEEE 300-bus power system that is disaggregated into 4 grid units (G^1, G^2, G^3 and G^4) as described in Chapter 5. An integration test of the load forecasting framework into the DMPC problem of Chapter 5 is conducted as follows. Our proposition is that a more accurate load forecasting can

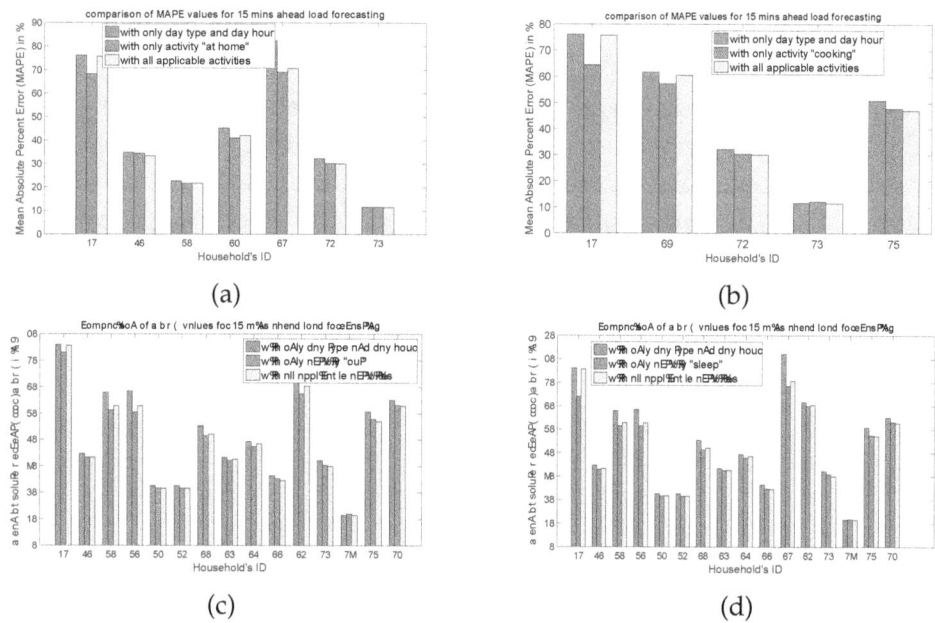

Figure 68.: MAPE comparison of 15 minutes ahead load forecasting influenced by 4 individual activities for certain households: a) "at home"; b) "cooking"; c) "out"; d) "sleep"

benefit the DMPC problem with a further cost reduction and stabilization. In this integration test, there is no available activity information. Therefore, an load forecasting for the DMPC problem is conducted by our load forecasting framework with only past load data as inputs.

As described in both Chapter 4 and Chapter 5, for the price determination and stabilization, forecasted nodal loads $\hat{d}$ are taken into account regarding the prediction horizon T. In the following integration test, we compare the optimization results of the DMPC problem with assumed future nodal loads and forecasted future nodal loads. The assumed future nodal loads are calculated based on the stochastic load model introduced in the evaluation setting of the previous chapters, while the forecasted future nodal loads are determined by a SVR-based load forecasting model that is trained with the same stochastic load model. For each bus node, an individual SVR forecasting model is trained.

In the case of forecasted nodal loads, we assume a homogenous load distribution on the available generators, which means that the power dispatch of the forecasted nodal loads for the prediction horizon T results in not only a matrix of forecasted nodal loads $\tilde{\hat{d}}(k) \equiv [(\hat{d}(k))^{tr}, \ldots, (\hat{d}(k+T-1))^{tr}]^{tr}$ but also a matrix of forecasted nodal generations $\tilde{\hat{s}}(k) \equiv [(\hat{s}(k))^{tr}, \ldots, (\hat{s}(k+T-1))^{tr}]^{tr}$. Then, based on the nodal power conservation expression in (26), a matrix of nodal power injections can be predicted: $\hat{\tilde{P}}_{in}(k) \equiv [(\hat{\tilde{P}}_{in}(k))^{tr}, \ldots, (\hat{\tilde{P}}_{in}(k+T-1))^{tr}]^{tr}$. Once the predictions of nodal loads, nodal generations and nodal injections are given, the DMPC problem in (54) can be solved under the constraints that are fortified with the predictions. Given the control variables $\Delta \bar{d}(k)$ and $\Delta \bar{s}(k)$, the constraints in (28), (29), (32), (56), (57) and (58) are enhanced with the predictions accordingly.

$$\Delta \bar{d}_i(k) \leq \hat{\tilde{d}}_i(k+1) - \bar{d}_i(k) \tag{76}$$

$$|\hat{\tilde{P}}_{i,in}(k) + \bar{\alpha}_i^{tr} \cdot (\Delta \bar{s}_i(k) + \Delta \bar{d}_i(k))| \leq f \cdot \bar{P}_{i,in}^{max} \tag{77}$$

$$\bar{r}_i^{down} \leq \hat{\tilde{s}}_i(k+1) - \hat{\tilde{s}}_i(k) \leq \bar{r}_i^{up} \tag{78}$$

$$\bar{s}_i^{min} \leq \hat{\tilde{s}}_i(k) \leq \bar{s}_i^{max} \tag{79}$$

$$\bar{s}_i^{min} \leq \hat{\tilde{s}}_i(k) + \Delta \bar{s}_i(k) \leq \bar{s}_i^{max} \tag{80}$$

$$\bar{P}_{i,in}^{min} \leq \hat{\tilde{P}}_{i,in}(k) \leq \bar{P}_{i,in}^{max} \tag{81}$$

$$\tag{82}$$

Then, we simulate the DMPC problem of the 4 interconnected grid units with two future nodal load cases, i.e. 1) with assumed nodal loads and 2) with forecasted nodal loads and enhanced constraints. In both cases, the simulation setting refers to the same configuration that is described in Chapter 4 and Chapter 5. The optimization goal for both cases is set equally that the nodal load adjustments Δd for all 4 grid units should be stabilized at the zero range.

Similar to the simulation setting in the previous chapters, simulations with 200 time steps for the local market price π of G^1, G^2, G^3 and G^4 under an agent cooperation are shown in Figure 69, Figure 70 and Figure 71, which refer to the optimization results with 3 different prediction horizon cases, i.e. $T = 1$, $T = 5$ and $T = 10$, respectively.

For all 3 prediction horizon cases, the simulation results with assumed nodal loads are considered as the baseline that represents the optimization results with 100% availability of the future load information. In case $T = 1$, see Figure 69, the optimized local market price π determined by forecasted nodal loads (in green dashed line) differs significantly from the baseline in red solid line for all 4 grid units. In comparison to case $T = 1$, Figure 70 for case $T = 5$ shows better convergence for all 4 grid units. Finally, we notice that in case $T = 10$ the optimized local market price π determined by forecasted nodal loads (in green dashed line) meets exactly the baseline in red solid line for the first 3 grid units G^1, G^2 and G^3, while the local market price π for the grid unit G^4 is although stabilized at 40 (monetary units) but diverges completely from the baseline (see Figure 71d). Overall, the simulation results confirm that our proposed load forecasting framework provides a higher accuracy for the greater prediction horizon, since the greater the prediction horizon is, the more the load data are taken into account for training the forecasting model. Moreover, in comparison to the baselines, a more accurate load forecasting contributes to the optimization results of the proposed DMPC problem, which concludes our above proposition.

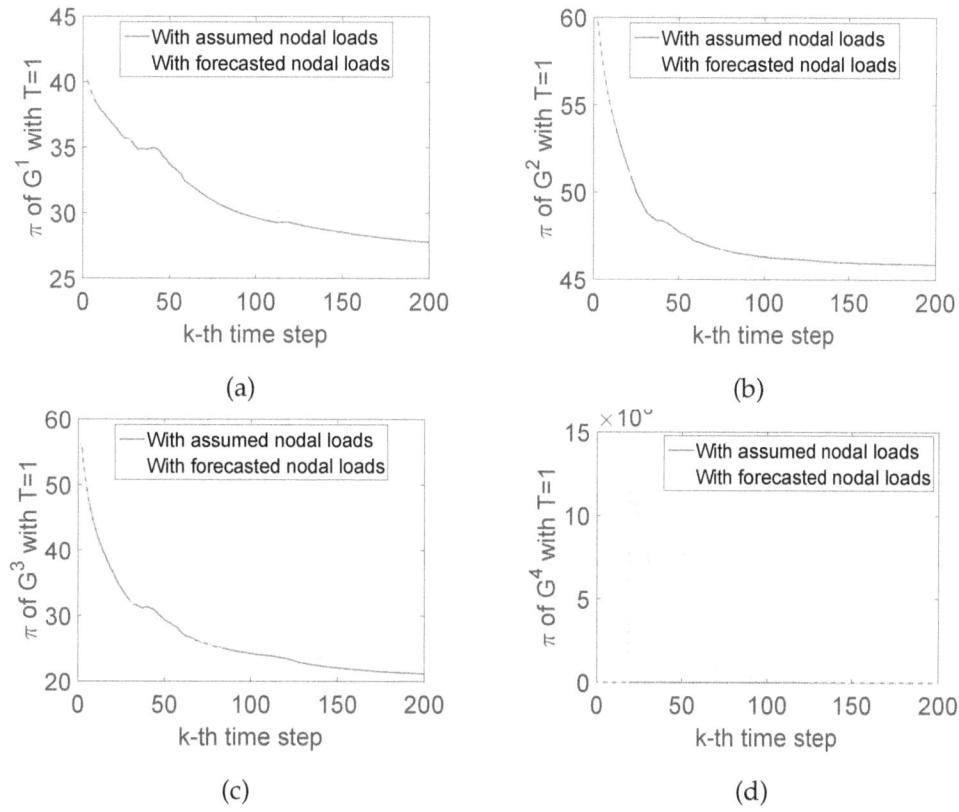

Figure 69.: Evolution of local market prices π with $T = 1$ of all 4 grid units with an agent cooperation: G^1 ((a) upper left), G^2 ((b) upper right), G^3 ((c) bottom left) and G^4 ((d) bottom right); The optimization results with assumed nodal loads and forecasted nodal loads are depicted in red solid line and green dashed line, respectively

6.8 CONCLUSION

In this chapter, an adaptive STLF framework was presented, which serves as a basis for real-time execution and evaluation of different activity recognition and load forecasting methods. The framework focuses on the algorithm implementation for activity recognition and load forecasting, and leaves out the need for developing sensor management, data persistence and data retrieval infrastructures as well as activity recognition and load forecasting controllers. Combining different influ-

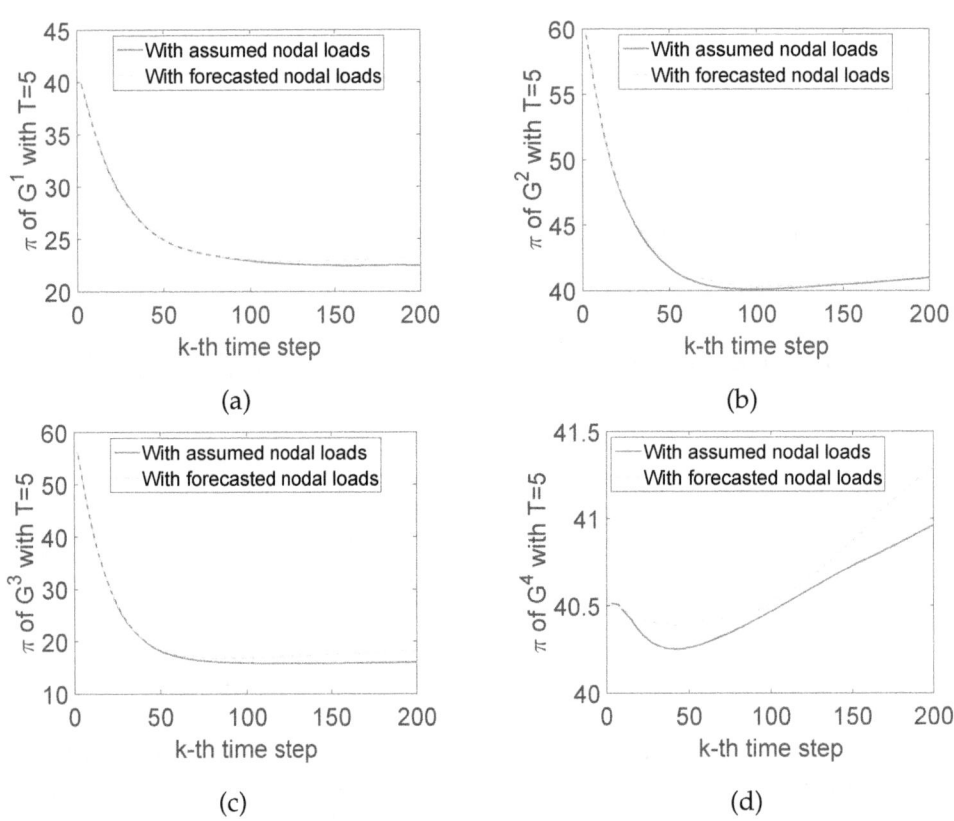

Figure 70.: Evolution of local market prices π with $T = 5$ of all 4 grid units with an agent cooperation: G^1 ((a) upper left), G^2 ((b) upper right), G^3 ((c) bottom left) and G^4 ((d) bottom right); The optimization results with assumed nodal loads and forecasted nodal loads are depicted in red solid line and green dashed line, respectively

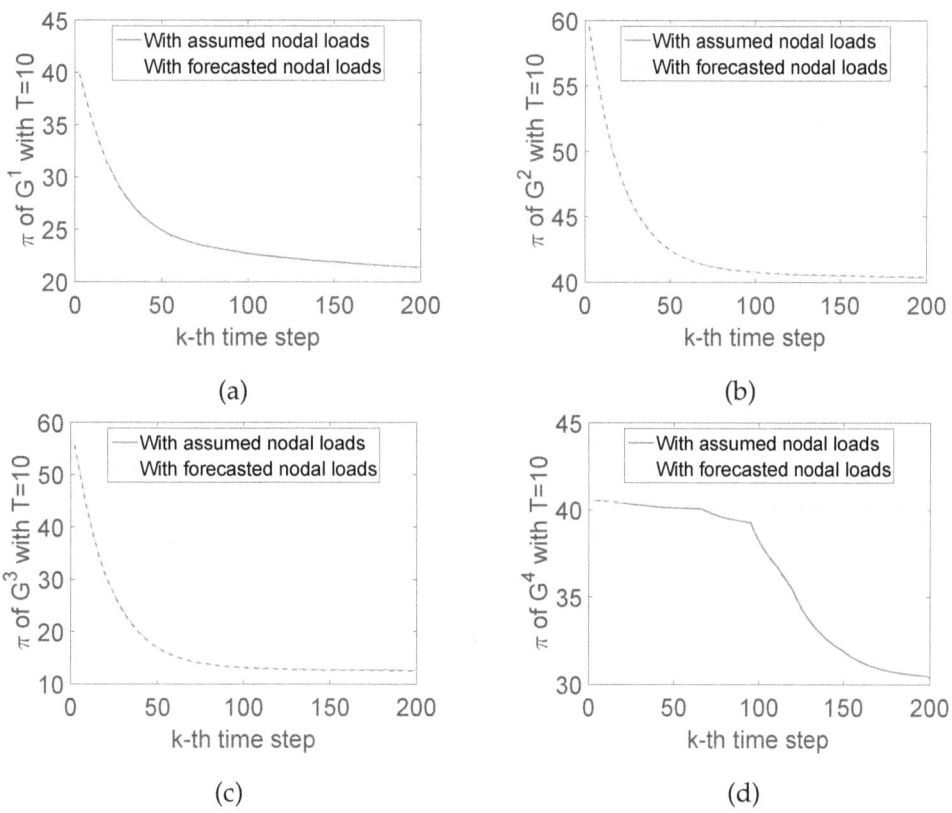

Figure 71.: Evolution of local market prices π with $T = 10$ of all 4 grid units with an agent cooperation: G^1 ((a) upper left), G^2 ((b) upper right), G^3 ((c) bottom left) and G^4 ((d) bottom right); The optimization results with assumed nodal loads and forecasted nodal loads are depicted in red solid line and green dashed line, respectively

encing factors, data preprocessing methods as well as forecasting algorithms, we conducted 3 different load forecasting case studies with 3 different datasets to show the impact of human activities. The experiment results are 3-fold:

1. A load forecasting at appliance level is in any case difficult due to the strong irregularity of the appliance-level power consumption. The strong irregularity implies as well that relevant human activities to the correspondent appliance-level power consumption are difficult to identify and thereby recognize.

2. Big events — the aggregation level of human activity data — have no impact on load forecasting at the individual meter level, but at the aggregation level.

3. Daily activity information as input variable could improve the accuracy of shorter term (e.g. 15 minutes ahead) household load forecasting. However, the contribution of individual activities for forecasting accuracy is household dependent and household specific.

Moreover, we integrated the proposed load forecasting framework into the DMPC problem of Chapter 5 for the IEEE 300 bus test case. The simulation results indicated that the load forecasting framework as an integrated process of the DMPC problem can provide accurate future load information for achieving the same optimization results regarding the market price as stated by the baseline with the assumption that the future load information is available.

Bibliography

[1] N. Amjady. Short-term bus load forecasting of power systems by a new hybrid method. *Power Systems, IEEE Transactions on*, 22(1):333–341, Feb 2007.

[2] Z.A. Bashir and M.E. El-Hawary. Applying wavelets to short-term load forecasting using pso-based neural networks. *Power Systems, IEEE Transactions on*, 24(1):20 –27, feb. 2009.

[3] M. Basu. Hybridization of bee colony optimization and sequential quadratic programming for dynamic economic dispatch. *International Journal of Electrical Power & Energy Systems*, 44(1):591 – 596, 2013. ISSN 0142-0615. doi: http://dx.doi.org/10.1016/j.ijepes.2012.08.026.

[4] Thomas M Cover and Joy A Thomas. *Elements of information theory*. John Wiley & Sons, 2012.

[5] Brandon Davito, Humayun Tai, and Robert Uhlaner. The smart grid and the promise of demand-side management. *McKinsey on Smart Grid*, pages 38–44, 2010.

[6] Yong Ding, Martin Alexander Neumann, Per Goncalves Da Silva, and Michael Beigl. A framework for short-term activity-aware load forecasting. In *Joint Proceedings of the Workshop on AI Problems and Approaches for Intelligent Environments and Workshop on Semantic Cities*, AIIP '13, pages 23–28, New York, NY, USA, 2013. ACM.

[7] Yong Ding, Martin A Neumann, Ömer Kehri, Geoff Ryder, Till Riedel, and Michael Beigl. From load forecasting to demand response-a web of things use case. In *Proceedings of the 5th International Workshop on Web of Things*, pages 28–33. ACM, 2014.

[8] Yong Ding, Julio Borges, Martin A. Neumann, and Michael Beigl. Sequential Pattern Mining - a Study to Understand Daily Activity Patterns for Load Forecasting Enhancement. In *1st IEEE International Smart Cities Conference (ISC2-2015)*, Guadalajara, Mexico, 2015. IEEE.

[9] Yong Ding, Sozo Inoue, Martin A. Neumann, Erwin Stamm, Xincheng Pan, and Michael Beigl. A Personalized Load Forecasting Enhanced by Activity Information. In *1st IEEE International Smart Cities Conference (ISC2-2015)*, Guadalajara, 2015. IEEE.

[10] Shu Fan and R.J. Hyndman. Short-term load forecasting based on a semi-parametric additive model. *Power Systems, IEEE Transactions on*, 27(1):134–141, Feb 2012. ISSN 0885-8950. doi: 10.1109/TPWRS.2011.2162082.

[11] A. Grandjean, J. Adnot, and G. Binet. A review and an analysis of the residential electric load curve models. *Renewable and Sustainable Energy Reviews*, 16(9):6539 – 6565, 2012. ISSN 1364-0321.

[12] R. Horta-Bernus, M. Rosas-Casals, and S. Valverde. Discerning electricity consumption patterns from urban allometric scaling. In *Complexity in Engineering, 2010. COMPENG '10.*, pages 49–51, 2010.

[13] Fahad Javed, Naveed Arshad, Fredrik Wallin, Iana Vassileva, and Erik Dahlquist. Forecasting for demand response in smart grids: An analysis on use of anthropologic and structural data and short term multiple loads forecasting. *Applied Energy*, 96(0):150 – 160, 2012. ISSN 0306-2619. doi: http://dx.doi.org/10.1016/j.apenergy.2012.02.027. Smart Grids.

[14] G.Y. Liu, Z.F. Yang, and B. Chen. Emergy-based urban dynamic modeling of long-run resource consumption, economic growth and environmental impact: conceptual considerations and calibration. *Procedia Environmental Sciences*, 13 (0):1179 – 1188, 2012. ISSN 1878-0296.

[15] Matteo Muratori, Matthew C. Roberts, Ramteen Sioshansi, Vincenzo Marano, and Giorgio Rizzoni. A highly resolved modeling technique to simulate residential power demand. *Applied Energy*, 107(0):465 – 473, 2013. ISSN 0306-2619.

[16] M. Mustermann, J. Smith, and A. Clark. A control loop approach for integrating the future decentralized power markets and grids. In *Smart Grid Communications (SmartGridComm), 2013 IEEE International Conference on*, pages 588–593, Oct 2013. doi: 10.1109/SmartGridComm.2013.6688022.

[17] Dongxiao Niu, Yongli Wang, and Desheng Dash Wu. Power load forecasting using support vector machine and ant colony optimization. *Expert Systems with Applications*, 37(3):2531 – 2539, 2010. ISSN 0957-4174.

[18] K. Nose-Filho, AD.P. Lotufo, and C.R. Minussi. Short-term multinodal load forecasting using a modified general regression neural network. *Power Delivery, IEEE Transactions on*, 26(4):2862–2869, Oct 2011.

[19] Abdeen Mustafa Omer. Energy, environment and sustainable development. *Renewable and Sustainable Energy Reviews*, 12(9):2265 – 2300, 2008. ISSN 1364-0321. doi: 10.1016/j.rser.2007.05.001. URL http://www.sciencedirect.com/science/article/pii/S1364032107000834.

[20] A.D. Papalexopoulos and T.C. Hesterberg. A regression-based approach to short-term system load forecasting. *Power Systems, IEEE Transactions on*, 5(4): 1535–1547, 1990.

[21] Carlos Ramos, Goreti Marreiros, Ricardo Santos, and Carlos Filipe Freitas. Smart offices and intelligent decision rooms. In *Handbook of Ambient Intelligence and Smart Environments*, pages 851–880. Springer, 2010.

[22] Xiaorong Sun, P.B. Luh, L.D. Michel, S. Corbo, K.W. Cheung, Wei Guan, and K. Chung. An efficient approach for short-term substation load forecasting. In *Power and Energy Society General Meeting (PES), 2013 IEEE*, pages 1–5, July 2013.

[23] Lukas G. Swan and V. Ismet Ugursal. Modeling of end-use energy consumption in the residential sector: A review of modeling techniques. *Renewable and Sustainable Energy Reviews*, 13(8):1819 – 1835, 2009. ISSN 1364-0321.

[24] FilipposS. Tymvios, SilasChr. Michaelides, and CharaS. Skouteli. Estimation of surface solar radiation with artificial neural networks. In Viorel Badescu, editor, *Modeling Solar Radiation at the Earth's Surface*, pages 221–256. Springer Berlin Heidelberg, 2008. ISBN 978-3-540-77454-9.

[25] Joakim Widn and Ewa Wckelgard. A high-resolution stochastic model of domestic activity patterns and electricity demand. *Applied Energy*, 87(6):1880 – 1892, 2010. ISSN 0306-2619.

7 | CONCLUSION AND OUTLOOK

A market-grid coupling represents an enhancement process of the real-time interaction between the market and the grid. More communication and interaction between the market and the grid offers great potential for grid relief and market price stabilization. A closed-loop feedback control system is used to interpret this interaction as control signals for a power dispatch control and a market price control. However, a feedback control loop usually refers to a unidirectional control problem that consists of a system plant and a controller. Modeling two control problems in a control loop for the market-grid coupling requires a fused problem formulation of a market price optimization and a power dispatch optimization towards an interoperable control. Therefore, the main contribution of this dissertation is the problem formulations, control strategies and numerical evaluations that are used to validate the feasibility of the proposed market-grid coupling with the interoperable controllability.

In Chapter 3, an intra-minute market model designed for a real-time price settlement as a balancing option was presented; a two-layer grid model was proposed for an optimal dynamic dispatch (ODD) study. Based on both models, a formal definition of the market-grid coupling was made by means of a feedback control loop. From the definitions and formulations as well as an initial test in this chapter, we can summarize the following findings:

- The market, the grid as well as the market-grid coupling all should work on a time scale of minutes.

- LMP-based nodal prices can be used as feedback signals to bridge the gap between the market and the grid.

- A real-time adaptation of the power market clearing based on the grid optimization output is feasible.

As a next step in Chapter 4, we introduced a further investigation and analysis of this formalized market-grid coupling in two different directions; i.e. a co-simulation framework and a MPC-based closed-loop feedback control system. The problem formulation of the control system firstly focuses on a coupling model with a single grid unit and its correspondent local market. Further simulation results confirmed the feasibility of the proposed market-grid coupling, from which we gained the following insights into the co-simulation framework and the centralized MPC problem formulation:

- With GridLAB-D and AMES as grid and market simulation components, it is possible to implement a co-simulation framework for the market-grid coupling.

- The memory usage of the co-simulation framework is highly dependent on the complexity of the simulated grid model.

- The convergent run time indicates a good scalability of the co-simulation framework.

- A nodal price analysis based on the co-simulation framework confirms again that LMP-based nodal prices can be used to couple the market with the grid.

- In the current version of the co-simulation framework, the proposed market-grid coupling has been only analyzed in terms of grid impact on the market price.

- The system modeling of a MPC-based closed-loop feedback control system enables an interoperable control between the market and the grid, in which a market price optimization and a power dispatch optimization are performed concurrently.

- The centralized MPC problem for a local market-grid coupling provides a market price stability and a power dispatch stability to varying degrees.

- The optimization based on the proposed control system meets the time response requirement (within a time scale of minutes) of the proposed market-grid coupling.

In Chapter 5, a distributed market-grid coupling model was developed and evaluated. We presented a distributed control architecture by means of a hierarchical MAS for extending the centralized MPC problem to a distributed MPC problem. A distributed MPC strategy was adopted to decompose the overall grid into interconnected grid units, so that individual grid units can achieve control objectives collaboratively. From a numerical study with the IEEE 300 bus test case that was disaggregated into 4 grid units, the following lessons were learned:

- The interconnected grid units can profit from the proposed collaborative MPC feedback loops for stabilizing their market price and power load dispatch concurrently.

- There is difficult for an isolated grid unit to stabilize either the load dispatch or the local market price without any inter-transmission.

- The more future information the MPC perceives, the higher stability the optimization results achieve.

- The execution time of the collaborative control loops with the DMPC agents is less than 70 s, which confirms that the proposed DMPC problem for each time step can be solved with a time scale of minutes as the response time requirement of the proposed market-grid coupling.

- The proposed MPC-based feedback control system can couple the market with the grid as two control problems in a control loop, and an optimal dynamic dispatch stabilizes the market price of individual grid units.

Finally, we presented an activity-aware load forecasting framework in Chapter 6, which was integrated in the previous DMPC problem formulation. Combining different influencing factors, data preprocessing methods as well as forecasting algorithms, we conducted 3 different load forecasting case studies with 3 different datasets to show the impact of human activities, and an integration test of the load forecasting framework into the DMPC problem. Gained insights from the framework design and the experiment results are 5-fold:

1. The load forecasting framework serves as a basis for real-time execution and evaluation of different activity recognition and load forecasting methods only by implementing the algorithms, leaving out the need for developing sensor management, data persistence and data retrieval infrastructures as well as activity recognition and load forecasting controllers.

2. A load forecasting at appliance level is in any case difficult due to the strong irregularity of the appliance-level power consumption. The strong irregularity also implies that relevant human activities to the correspondent appliance-level power consumption are difficult to identify and thereby recognize.

3. Big events — the aggregation level of human activity data — have no impact on load forecasting at the individual meter level, but at the aggregation level.

4. Daily activity information as input variable could improve the accuracy of shorter term (e.g. 15 minutes ahead) household load forecasting. However, the contribution of individual activities for forecasting accuracy is household dependent and household specific.

5. The load forecasting framework as an integrated process of the DMPC problem can provide accurate future load information for achieving the same optimization results as the baseline determined with the assumption that the future load information is available.

This dissertation presented a novel work to formulate the market-grid coupling as an interoperable control problem. With a MPC-based approach, a market price optimization problem and a power dispatch optimization problem can be coupled in one control loop. With a distributed control strategy, the centralized MPC problem can be extended to a distributed MPC problem for interconnected grid units. The numerical studies based on the IEEE bus system test cases confirmed our proposition that a MPC-based feedback control system can couple the market with the grid as two control problems in a control loop, and an optimal dynamic dispatch stabilizes the market price.

Since the stability of the proposed coupling system has been evaluated mainly based on numerical simulations, a control-theoretical stability analysis on this market-grid coupling model can be considered as future work, in order to theoretically confirm the ability of the proposed feedback control system for maintaining the steady-state operation of both the power market and the power grid under physical disturbances and constraints of the power system. Furthermore, the evaluation of this work is mainly based on synthetic grid models and several simplifications regarding load models, a further evaluation with real-world grid data and more realistic load/consumption models is required as future work as well. Regarding pricing functions, we have introduced two price stabilizing functions determined by the λ-based method and the subgradient-based method. However, only the λ-based method was applied in the evaluation phase. Therefore, a further study on the subgradient-based method as well as other dynamic pricing mechanisms for stabilizing an optimal power dispatch provides another future research direction.

List of Symbols and Abbreviations

A

AI   Artificial Intelligence.

AMI   Advanced Metering Infrastructure.

ANN   Artificial Neural Network.

AR   Auto Regression.

ARIMA   Autoregressive Integrated Moving Average.

ARMAX   Autoregressive Moving Average Model with Exogenous Inputs.

B

BMBF   Federal Ministry of Education and Research.

BRP   Balance Responsible Party.

BVP   Boundary Value Problem.

C

CA   Call Auction.

CDA   Continuous-Double-Auction.

CET   Central European Time.

CSF   Conjectured Supply Function.

CSS   Cascading Style Sheets.

CWE   Central Western Europe.

D

DA   Double Auction.

DAE   Differential Algebraic Equation.

DED   Dynamic Economic Dispatch.

DEMD   Differential Empirical Mode Decomposition.

DER   Distributed Energy Resources.

DG   Distributed Generation.

DLMP   Dynamic Locational Marginal Pricing.

DMPC   Distributed Model Predictive Control.

DR   Demand Response.

DSM   Demand Side Management.

DSO   Distribution System Operator.

E

EEG   Rnewable Energy Law in Germany.

EEX   European Energy Exchange.

EP   Equilibrium Point.

G

GCP   Grey Correlation Projection.

GenCos   Generation Companies.

GLM   refers to the GridLAB-D model/configuration file.

H

HTML   HyperText Markup Language.

I

inter-grid   refers to anything between grid units.

intra-grid   refers to anything within a grid unit.

IoE   Internet of Energy.

IoT   Internet of Things.

ISO   Independent System Operator.

J

JSON   JavaScript Object Notation.

K

KKT   KarushKuhnTucker.

L

LMP   Locational Marginal Price/Pricing.

LQR   Linear-Quadratic Regulator.

LS-SVM   Least-Squares Support Vector Machine.

LSEs   Load Serving Entities.

M

MAPE   Mean Absolute Percentage Error.

MAS   Multi-Agent System.

MGCC   MicroGrid Central Controller.

MIQP   Mixed-Integer Quadratic Programming.

MPC   Model Predictive Control.

MPEC   Mathematical Program with Equilibrium Constraints.

MSE   Mean Square Error.

LIST OF SYMBOLS AND ABBREVIATIONS

N

NOBEL   refers to the EU FP7 project: Neighbourhood Oriented Brokerage Electricity and monitoring system.

O

OCDD   Optimal Control Dynamic Dispatch.

ODD   Optimal Dynamic Dispatch.

OPF   Optimal Power Flow.

P

PID   Proportional-Integral-Derivative.

PNNL   Pacific Northwest National Laboratory.

PSAT   Power System Analysis Toolbox.

PSO   Particle Swarm Optimization.

Q

QuadProgJ   refers to an open-source Java optimization solver.

R

RD   Replicator Dynamics.

REST   Representational State Transfer.

RF   Random Forest.

RNN   Recurrent Neural Network.

RTP   Real-Time Pricing.

S

SDP   Semidefinite Programming.

SFE   Supply Function Equilibrium.

SOC   State of Charge.

SOM   Self-Organized Map.

SR   Spinning Reserve.

std   Standard Deviation.

STLF   Short-Term Load Forecasting.

SVM   Support Vector Machine.

SVR   Support Vector Regression.

Symbol A denotes the activity information.

Symbol A^i denotes a weighted adjacency matrix that refers to the nodal admittance matrix.

Symbol a_{kk}^i denotes the self-admittance at node v_k^i.

Symbol a_{kl}^i denotes the admittance of the branch e_{kl}^i.

Symbol $\alpha_{0,1,2}^i$ denotes the cost coefficients for polynomial cost functions.

Symbol α_{ij} denotes a sensitivity factor for strengthening the transmission limitation.

Symbol b denotes a penalization constant for the imbalance of the deviation of the inter-transmission power flows of one grid unit between the current optimization iteration s and the previous one $s-1$.

Symbol $\bar{\lambda}_{i \leftarrow j}(k|s)$ denotes the Lagrange multipliers in interconnection term of the DMPC objective function.

Symbol $\bar{\lambda}_{i \rightarrow j}(k|s)$ denotes the Lagrange multipliers in interconnection term of the DMPC objective function.

Symbol $\bar{\pi}_{i,g}(k)$ denotes the nodal local market prices for the generation power.

Symbol $\bar{\pi}_{i,in}(k)$ denotes the "toll charges" for the injected power as revenue of network players.

Symbol $\bar{\pi}_{i \leftarrow j}(k)$ denotes the local market prices of the grid unit G^j.

Symbol $\bar{\pi}_{i,l}(k)$ denotes the nodal utility multipliers for the load power.

LIST OF SYMBOLS AND ABBREVIATIONS

Symbol $\bar{P}_{i,in}$ denotes the output vector representing nodal power injections for the grid unit G^i.

Symbol $\bar{P}_{i,in}^{min/max}$ denotes the upper and lower bounds of the intra-transmission power flow.

Symbol $\bar{\pi}_{i \to j}(k)$ denotes the local market prices of the grid unit G^i.

Symbol $\bar{r}_i^{down/up}$ denotes the down and up ramp rates of generators.

Symbol $\bar{s}_i^{min/max}$ denotes the upper and lower bounds of the nodal generation.

Symbol $\bar{w}_{i,in}$ denotes the vector of interconnection input power flows for the grid unit G^i.

Symbol $\bar{w}_{i,in/out}^{min/max}$ denotes the intra- and inter-transmission power flow limits.

Symbol $\bar{w}_{i \leftarrow j,in}$ denotes the inter-transmission power flow input to the grid unit G^i from G^j.

Symbol $\bar{w}_{i,out}$ denotes the vector of interconnection output power flows for the grid unit G^i.

Symbol $\bar{w}_{i \to j,out}$ denotes the inter-transmission power flow output from the grid unit G^i to G^j.

Symbol β^j denotes the different consumer/load types.

Symbol $\bullet|s$ denotes the current optimization iteration.

Symbol C denotes a dynamic coefficient matrix.

Symbol c denotes a penalization constant for the imbalance of the inter-transmission power flows between two interconnected grid units.

Symbol c_i denotes a cost function regarding the total generation cost at node i.

Symbol c_i^P denotes a cost function regarding the generation cost for injected power at producer node i.

Symbol c_i^R denotes a cost function regarding the generation cost for reserve power at producer node i.

Symbol c_j denotes a disutility function at consumer node j.

Symbol D denotes the day type information for a load forecasting.

212

Symbol d denotes the demand vector representing complex nodal power consumption.

Symbol $\Delta \bar{d}_i$ denotes the control vector representing nodal load adjustments for the grid unit G^i.

Symbol $\Delta \bar{s}_i$ denotes the control vector representing nodal generation adjustments for the grid unit G^i.

Symbol $d \in \mathbb{D}$ denotes a consumer node.

Symbol d_j denotes the power consumption or the inverse of W_j at bus node j.

Symbol $D_j(t)$ denotes power demand at consumer or prosumer node j for time t.

Symbol $EF_x(t)$ denotes the external factors for a load forecasting.

Symbol E^i denotes a set of directed edges of the graph G^i, which refers to the set of intra-transmission branches.

Symbol E^{ij} denotes the set of inter-transmission branches between the grid unit G^i and G^j.

Symbol e_{kl}^i denotes a directed edge that represents the branch, in which power flows from node v_k^i to node v_l^i.

Symbol ϵ_j denotes an additive white noise.

Symbol f denotes a percentage parameter for strengthening the transmission limitation.

Symbol $f(\bullet)$ denotes the common system equation of a state-space representation.

Symbol $f_d(P_d)$ denotes a general form of cost functions of consumer powers.

Symbol $f_p(P_p)$ denotes a general form of cost functions of prosumer powers.

Symbol $f_s(P_s)$ denotes a general form of cost functions of generator powers.

Symbol γ denotes the weighting factor of the λ-based method.

Symbol $g(\bullet)$ denotes the common system equation of a state-space representation.

Symbol G^i denotes a directed graph i.

Symbol g_P denotes the set of non-linear nodal power balance equations for real powers.

Symbol g_Q denotes the set of non-linear nodal power balance equations for reactive powers.

Symbol H denotes the day hour information for a load forecasting.

Symbol $\hat{d}(k)$ denotes the forecasted nodal loads at time step k.

Symbol $H(P)$ denotes the entropy expression.

Symbol $H(P|\bullet)$ denotes the conditional entropy expression.

Symbol $h(x)$ denotes a vector of all the inequality transmission constraints.

Symbol I^i denotes a complex vector of node current injections of the grid unit G^i.

Symbol $I(P; A)$ denotes the mutual information expression.

Symbol $I(P; D, H)$ denotes the mutual information expression.

Symbol J denotes a cost function.

Symbol $j(\bullet)$ denotes the cost function at each time step.

Symbol k denotes the discrete time step.

Symbol λ denotes a nodal price vector or a Lagrange multiplier.

Symbol $\lambda_n(t)$ denotes a nodal price at bus node n for time t.

Symbol $LF(\bullet)$ denotes the general form of a load forecasting function.

Symbol l_P denotes the path length that refers to the number of edges in the path P.

Symbol l_P^w denotes the weighted path length.

Symbol M refers to the number of transmission lines or branches.

Symbol $\mathbb{D}$ denotes the set of consumers.

Symbol $\mathbb{L}$ denotes the Lagrange dual function associated to the OPF problem.

Symbol $\mathbb{N}$ denotes the set of all natural numbers without 0.

Symbol $\mathbb{N}_0$ denotes the set of all natural numbers with 0.

Symbol $\mathbb{R}$ denotes the set of real numbers.

Symbol $\mathbb{S}$ denotes the set of producers.

Symbol $\mathbb{T}$ denotes the set of network players.

Symbol $\mathcal{A}$ denotes the set of meta activities.

Symbol $\mathcal{N}$ denotes the symbol of a normal distribution.

Symbol $\mathcal{G}(\pi(k))$ denotes the subgradient direction regarding λ in the LMP formulation.

Symbol $\mathcal{J}$ denotes the Jacobian matrix.

Symbol $\mathcal{J}_m$ denotes the Jacobian matrix.

Symbol $\mathcal{L}$ denotes the Lagrange dual function of the OPF problem.

Symbol μ_{max} denotes barrier parameters for the logarithmic barrier function of the slack variables in the Lagrange dual function $\mathbb{L}$.

Symbol μ_{min} denotes barrier parameters for the logarithmic barrier function of the slack variables in the Lagrange dual function $\mathbb{L}$.

Symbol N refers to the number of bus nodes.

Symbol N_g denotes the number of grid units.

Symbol Ω_{G^i} denotes the set of interconnections to and from G^i.

Symbol Ω_l^i denotes the set of bus nodes connected to node v_l^i within G^i.

Symbol P denotes a path of the directed graph G.

Symbol $P_a(i)$ denotes the time series of the actual power load.

Symbol $pdf(\bullet)$ denotes the expression of a probability density function.

Symbol $P_{D_j}(t)$ denotes consumed power at consumer or prosumer node j for time t.

Symbol $\Phi(\bullet)$ denotes the objective function of the LMP formulation.

Symbol $\Phi_{i,inter|j}(\bullet)$ denotes a grid interconnection sub-function.

Symbol ϕ_{ij} denotes power flow limits of the active powers flowing through the branches from node i to j.

Symbol $\Phi_{i,local}(\bullet)$ denotes a local grid objective function.

Symbol π denotes a local market price vector.

Symbol $P_{i,in}$ denotes the nodal injection real power at node i.

Symbol Π_k denotes the general form of a pricing function.

Symbol $\Pi^\lambda(\bullet)$ denotes the price stabilizing function of the λ-based method.

Symbol $\Pi^\mathcal{G}(\bullet)$ denotes the price stabilizing function of the subgradient-based method.

Symbol P_{in}^i denotes the active power flows into the grid unit G^i.

Symbol $P_{in,j}^i$ denotes the active power flow into the grid unit G^i through the node j.

Symbol $P_{in}^l(t)$ denotes power injection at bus node l for time t.

Symbol $p \in \mathbb{P}$ denotes a prosumer node.

Symbol $\pi_n(t)$ denotes a local market price at bus node n for time t.

Symbol $pmf(\bullet)$ denotes the expression of a probability mass function.

Symbol $P_{n,g}^i$ denotes the active generator power at bus node v_n^i.

Symbol $P_{n,in}^i$ denotes the active net power injection at bus node v_n^i.

Symbol $P_{n,l}^i$ denotes the active load power at bus node v_n^i.

Symbol P_{out}^i denotes the active power flows out of the grid unit G^i.

Symbol $P_{out,j}^i$ denotes the active power flow out of the grid unit G^i through the node j.

Symbol $P_p(i)$ denotes the time series of the predicted power load.

Symbol P_s denotes the shortest path.

Symbol $P_{S_i}(t)$ denotes power generation for consumers at generator or prosumer node i for time t.

Symbol $P(t-y)$ denotes the mean power load value of the previous time windows.

Symbol Q_i denotes the weighting matrix in the social welfare expression.

Symbol $Q_{i,in}$ denotes the nodal injection reactive power at node i.

Symbol Q_{in}^i denotes the reactive power flows into the grid unit G^i.

Symbol $Q_{n,g}^i$ denotes the reactive generator power at bus node v_n^i.

Symbol $Q_{n,in}^i$ denotes the reactive net power injection at bus node v_n^i.

Symbol $Q_{n,l}^i$ denotes the reactive load power at bus node v_n^i.

Symbol Q_{out}^i denotes the reactive power flows out of the grid unit G^i.

Symbol Q_{ui} denotes the weighting sub-matrix regarding the costs of the generation and load adjustment.

Symbol Q_{xi} denotes the weighting sub-matrix regarding the costs of the generation and load power.

Symbol Q_{yi} denotes the weighting sub-matrix regarding the costs of the power injection.

Symbol ρ_P denotes the Lagrange multipliers in the Lagrange dual function $\mathbb{L}$.

Symbol ρ_{Pr} denotes the Lagrange multipliers of the real power balance equation at the reference node (slack bus).

Symbol ρ_Q denotes the Lagrange multipliers in the Lagrange dual function $\mathbb{L}$.

Symbol ρ_{Qr} denotes the Lagrange multipliers of the reactive power balance equation at the reference node (slack bus).

Symbol R_i denotes the weighting matrix in the social welfare expression.

Symbol $R_{S_i}(t)$ denotes reserve generation for the balancing market at generator or prosumer node i for time t.

Symbol $r(t)$ denotes the reference value.

Symbol R_{ui} denotes the weighting sub-matrix regarding the costs of the generation and load adjustment.

Symbol R_{xi} denotes the weighting sub-matrix regarding the costs of the generation and load power.

Symbol R_{yi} denotes the weighting sub-matrix regarding the costs of the power injection.

Symbol s denotes also a vector of slack variables in the Lagrange dual function $\mathbb{L}$.

Symbol s denotes the supply vector representing complex nodal power generation.

Symbol s_i denotes the power generation or the inverse of W_i at bus node i.

Symbol σ_j denotes the standard deviation of the additive white noise ϵ_j.

Symbol σ_{kl} denotes the number of all shortest paths between the node v_k and v_l.

Symbol $\sigma_{kl}(e)$ denotes the number of shortest paths that pass through the edge e.

Symbol $s \in \mathbb{S}$ denotes a generator node.

Symbol T denotes the total time steps or the prediction horizon.

Symbol Θ^i denotes the vector of nodal voltage angles.

Symbol θ^i_{kl} denotes the phase angle difference between node v^i_k and v^i_l.

Symbol $\tilde{u}$ denotes variables over the prediction horizon T with the tilde symbol over the vector variables.

Symbol $\tilde{v}$ denotes the set of all optimization variables over the prediction horizon T.

Symbol $\tilde{\tilde{d}}(k)$ denotes a matrix of forecasted nodal loads.

Symbol $\tilde{\tilde{P}}_{in}(k)$ denotes a matrix of forecasted nodal power injections.

Symbol $\tilde{\tilde{s}}(k)$ denotes a matrix of forecasted nodal generations.

Symbol tr denotes a matrix transpose.

Symbol u_j denotes a utility function at consumer node j.

Symbol $u(t)$ denotes the control variable.

Symbol $\varphi(x)$ denotes an objective function of the optimization variable x.

Symbol V^i denotes also a complex vector of node voltages of the grid unit G^i.

Symbol V^i denotes a set of vertices of the graph G^i, which refers to the set of n_i bus nodes.

Symbol $W^{-1}(\bullet)$ denotes the inverse function of $W(\bullet)$ that refers to a price-to-consumption mapping function.

Symbol W_i denotes the welfare of a producer.

Symbol W_j denotes the welfare of a consumer.

Symbol W_l denotes the welfare of a network player.

Symbol $W_{tot}(t)$ denotes the total welfare at time t.

Symbol $x(t)$ denotes the state variable.

Symbol $y(t)$ denotes the controlled variable.

T

TOU   Time-Of-Use.

TSO   Transmission System Operator.

W

WSN   Wireless Sensor Network.

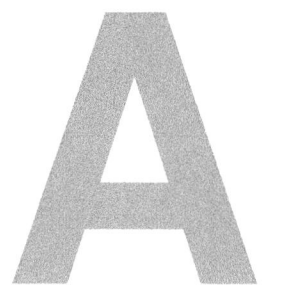

APPENDIX

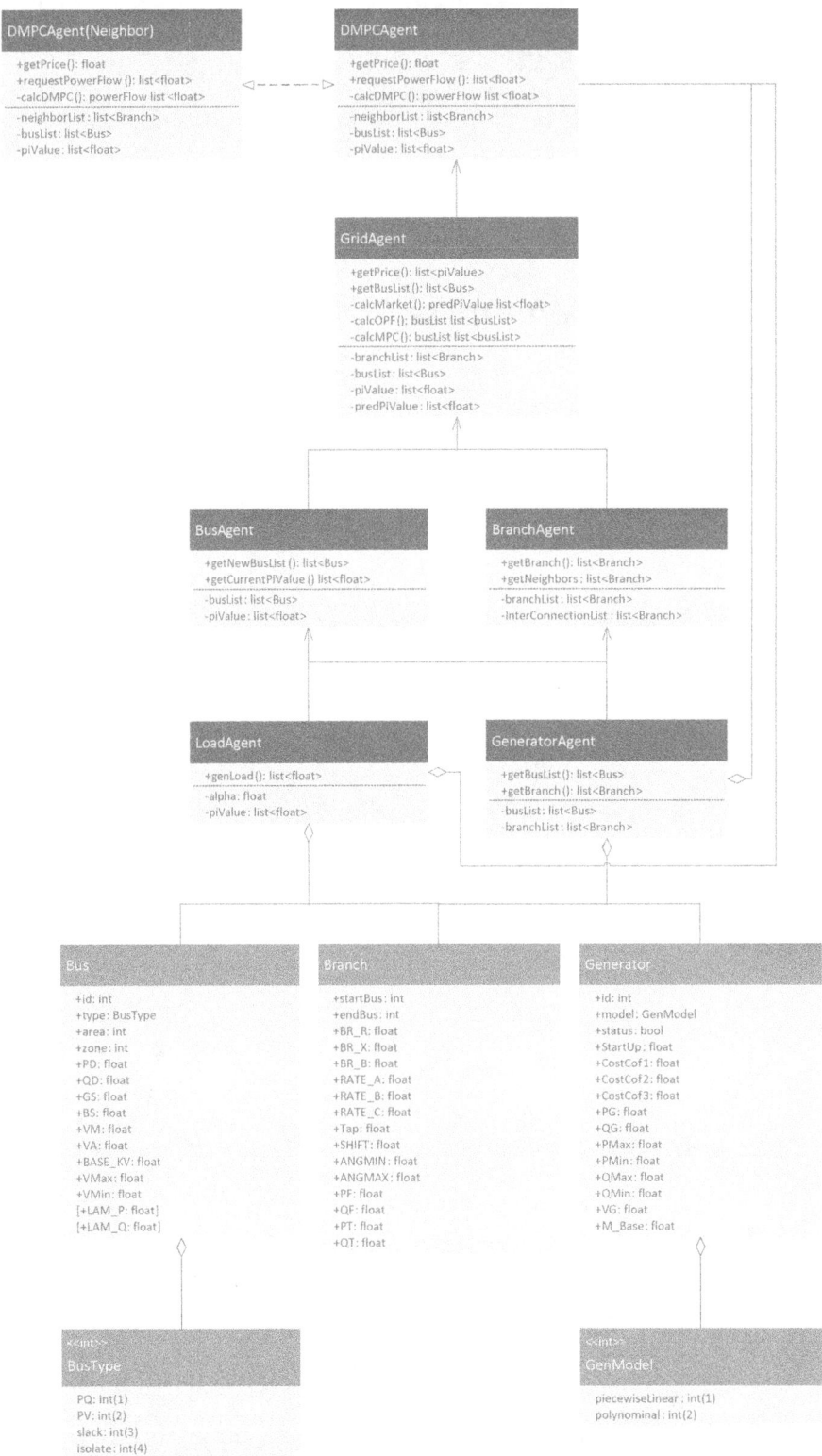

Figure 72.: The class diagram of the DMPC implementation with all agent classes

www.ingramcontent.com/pod-product-compliance
Lightning Source LLC
Chambersburg PA
CBHW080652190526

45169CB00006B/2089